Proceedings

of the

International Beilstein Workshop

MOLECULAR INFORMATICS:

CONFRONTING COMPLEXITY

May, 16th - 26th, 2002

Bozen, Italy

Edited by Martin G. Hicks and Carsten Kettner

BEILSTEIN-INSTITUT ZUR FÖRDERUNG DER CHEMISCHEN WISSENSCHAFTEN

Trakehner Str. 7 – 9
60487 Frankfurt
Germany

Telephone:	+49 (0)69 7167 3211	**E-mail:**	info@beilstein-institut.de
Fax:	+49 (0)69 7167 3219	**Web-page:**	www.beilstein-institut.de

IMPRESSUM

Molecular Informatics: Confronting Complexity, Martin G. Hicks and Carsten Kettner (Eds.), Proceedings of the Beilstein-Institut Workshop, May 13th - 16th, 2002, Bozen, Italy.

Bibliographic information published by *Die Deutsche Bibliothek*

Die Deutsche Bibliothek lists this publication in the *Deutsche Nationalbibliografie*; detailed bibliographic data is available in the internet at http://dnb.ddb.de

ISBN 3-8325-0319-6

Layout by: proLeaf
Schwerinstrasse 28
40477 Duesseldorf
Tel: +49 (0) 2 11 492 14 90
Fax: +49 (0) 2 11 492 14 91
Internet: http://www.proleaf.de

Printed by: Logos Verlag Berlin
Comeniushof, Gubener Str. 47
10243 Berlin
Tel: +49 (0) 30 42 85 10 90
Fax: +49 (0) 30 42 85 10 92
Internet: http://www.logos-verlag.de

PREFACE

The Beilstein Institute organizes and sponsors scientific meetings, workshops and seminars, with the aim of catalysing advances in chemical science by facilitating the interdisciplinary exchange and communication of ideas amongst the participants.

This workshop – *Molecular Informatics: Confronting Complexity* - addressed some of the new challenges that face scientists in the post-genome era, in particular, the integration of two, until recently, disparate sciences – chemistry and biology. The underlying theme of the workshop was to gain insight into the behaviour of biological and molecular systems through the application of molecular informatics.

The flood of data being generated as a result of research into genomics and proteomics is often overwhelming. Well publicised successes tend to draw the focus away from some of the significant issues relating to a better understanding of molecular systems which are still far from clear. Whereas the development of predictive models based on analogy has been very successful in chemistry and cheminformatics, the non-linear nature of biomolecular systems, often with multiple pathways, restricts similar transference within bioinformatics. However, without a critical analysis, taking into account the assumptions and limitations of hypotheses and predictive models, advances in molecular informatics will not assume significance. Before this can be effectively carried out, more effort needs to be made in bridging the gap between chemists, dealing with the structure and properties of molecules, and biologists, working with complex molecular and cell physiological systems.

Participants, as well as, speakers were confronted with the following complex challenges from cheminformatics and bioinformatics: knowledge discovery and data mining, rational drug design, prediction of small molecule bioavailability (ADME Tox) properties, protein structure and function determination, new methods of drug-target modeling, cellular metabolism and metabolic pathways, and the use of high-throughput methods (biochips, x-ray crystallography) for acquiring gene expression and protein structure, as well as, binding information.

This meeting did not set out to solve all problems, but to initiate a dialog between scientists of different disciplines. Over the three days of the workshop, the participants not only heard excellent talks, took part in lively discussions, but in the time between the official sessions of the scientific program, exchanged ideas and thoughts and generally made a valuable and personal contribution to bridging the gap!

We would like to thank particularly the authors who provided us with written versions of the papers that they presented. Special thanks go to all those involved with the preparation and organization of the workshop, to the chairmen who piloted us successfully through the sessions, and to the speakers and participants for their contribution in making this workshop a success.

Frankfurt / Main, May 2003

Martin G. Hicks
Carsten Kettner

CONTENTS

STUDIES ON YEAST MEMBRANE TRANSPORTERS – HOW CAN COMPUTATIONAL BIOLOGY HELP?

CARSTEN KETTNER

Beilstein-Institut zur Förderung der chemischen Wissenschaften,
60487 Frankfurt/Main, Germany
E-Mail: ckettner@beilstein-institut.de

Received: 12th June 2002 / Published: 15th May 2003

ABSTRACT

With the availability of complete genome sequences, emphasis has shifted toward the understanding of protein function and this in turn has opened up a new "-omics"-field, i.e. functional proteomics. Structural studies of proteins are only one aspect of functional proteomics and are mostly carried out by computational means. However, these investigations must be completed by function studies resulting in structure/function relationships and this can only be accomplished at the lab bench.

Some examples of comprehensive investigations on transport proteins of yeast, *Saccharomyces cerevisiae*, can be used to illustrate these relationships. This research includes various methods and tools concerning visualisation, sequencing and annotation. In addition, the transport activity of a tonoplast-residing proton pump has been studied in detail by biophysical approaches. The result of these investigations on the structure/function relationships demonstrate a fruitful cooperation of so-called traditional "wet" biology and computational biology.

INTRODUCTION

In scientific discussions the question is often raised whether bioinformatics and cheminformatics are equivalent or at least overlapping disciplines since computational scientists from both disciplines often seem to work in the same field. This in turn leads to a further question, i.e. if this is not true then is it possible to bridge or at least narrow the gap between them (1)?

However, if we take a step back and examine the situation we see that although computer scientists and biologists are accustomed to working together, they still think of their respective disciplines as separate. The use of computers as a new tool for investigation and research has reached the apparent traditionally techniques-free biology which is often described as a change of paradigm in biology and begs the question whether the biologists themselves are ready for the integration of the "classic" biology, as an experimental and practical discipline, and the computer-aided area in which bioinformatics plays a central role as part of the computational biology.

However, science without computers is unimaginable, since they have inserted themselves into almost every aspect of laboratory life, for example, for collecting, analysing and plotting experimental data, for aiding research topics by retrieving corresponding databases such as PubMed to search for literature or genomic or proteomics databases to look for successful research candidates, for the modelling of enzyme reaction cycles, enzyme-substrate interactions, protein foldings and 3D structures, and last but not least, for writing papers and grant applications.

One important aspect of the "ditch" between traditional biology and bioinformatics might be that there are a number of various online databases available which allows researchers to carry out their investigations and discoveries without even setting a foot in a lab.

Thus, if one accesses PubMed or PDB database (http://www.ncbi.nlm.nih.gov/entrez/query.fcgi?db=PubMed; http://www.rcsb.org/pdb/) to query the actual literature or any protein, one will recognize that there are numerous reports on bioinformatic-handled proteins concerning their sequence, modelled structure or even perhaps their subcellular location. Consequently, the least reports deal in fact with proven functional properties or regulatory aspects of these proteins. Shortfalls in the ability of bioinformatics to predict both the existence and function of genes have also illustrated the need for protein analysis which has given rise to a new research field, called proteomics. The emergence of proteomics has been inspired by the realization that the final product of a gene is inherently more complex and closer to function than the gene itself. The most practical application of proteomics is the analysis of single proteins as opposed to entire proteomes. This type of proteomics, which is referred to as "functional proteomics", is always driven by a specific biological question and requires a huge arsenal of both experimental methods and techniques as well as computational approaches to model, predict and explain biological function at the molecular level. Thus, the combination of protein identification by bioinformatics and characterization by "wet" biology has a meaningful outcome. Consequently, this is an unique chance to bridge the gap between the number of gene sequences in databases and the number of functionally characterized gene products which is currently a major challenge in biology.

As one example for the successful conquest of the scientific impediments, an important model organism for functional proteomics is presented, a short overview about both transport mechanisms across one biological membrane (the tonoplast) and some insight to a powerful method – the patch clamp technique – for the functional characterization of one representative of ion transporting proteins, which is an ATP-driven proton pump, is given. At the same time, the utilization of computers as an equally powerful tool for the collection, analysis and presentation of experimental data is demonstrated.

THE TEST ORGANISM IS AN IMPORTANT MODEL ORGANISM

Rather than isolate the pump for studies in an artificial environment, the enzyme was investigated in its "quasi" native environment. In keeping with tradition, the fungus baker's yeast *Saccharomyces cerevisiae* was chosen for this purpose. Since the mid-1980s, an ever-increasing number of molecular biologists and physiologists have used yeast as their primary research system and consequently, this has resulted in a virtually autocatalytic stimulus for continuing investigations of all aspects of molecular and cell biology. The "awesome power of yeast genetics" has become legendary and is the envy of those who work with higher eukaroytes. The complete sequence, published in 1996 and containing ca. 12 kb of DNA packed into 16 chromosomes with 6300 genes identified (2, 3), has proved to be extremely useful as reference for the sequencing of human and other higher eukaryotic genes. For example, of the 80 human disease genes so far identified, 12 yeast homologues have been found (4). Comparative studies have resulted in the suggestion that most basic biological functions of eukaryotic cells are carried out by a core set of orthologous "house-keeping" proteins. Thus, the assumption that analysis of yeast proteins will give insight into those of higher eukaryotes is valid.

Furthermore, this model organism provides a series of further advantages. Yeast is a free-living, unicellular eukaryote. It is the best characterized genetic system which makes it easily available for genetic manipulation and can be used for conveniently analysing and functionally dissecting gene products from itself and other eukaryotes. Yeast cells are highly versatile DNA transformation systems and viable with numerous markers. They are very suitable for replica plating and mutant isolation and easy to handle in batch cultures in which cells reach stationary phase after 18 to 22h.

For researchers who are mostly interested in transport processes across the membrane, some additional features make yeast a very interesting organism. Yeast cells facilitate structure-function studies of any electrogenic or electrophoretic ion transporters which can be expressed in the plasma membrane or tonoplast. The membranes themselves act as an ample source of diverse membrane proteins, such as ion channels, pumps and cotransporters, which lend themselves to electrophysiological analysis and specifically to patch-clamping. There are currently about 258 recognized and putative transporters from bioinformatic studies within the genome of *S. cerevisiae* but only perhaps one dozen of these transport proteins have been so far functionally characterised.

Subcellular Location of the Pump

The subcellular location of the focused proton pump is the vacuolar membrane. Plant and fungal vacuoles are intracellular compartments, bordered by the tonoplast, the vacuolar membrane, and occupy up to 90% of the intracellular volume (Fig. 1). They are multifunctional organelles with specific properties which are central to the cellular strategies of development of plants and fungi and, furthermore, as an acidic compartment they share some of their fundamental properties with mammalian lysosomes. Many newly synthesized proteins are targeted to the vacuole through the secretory pathway, and there they undergo maturing, processing and sorting processes (5). They confer the ability to accumulate a wide variety of solutes to relatively high concentrations and separate these solutes from the mainstream metabolism. Vacuoles store diverse metabolites, such as carbohydrates, amino acids, organic acids as well as inorganic phosphate, sulphate, calcium, potassium, sodium and other ions (6, 7). Here, degradation of carbohydrates and peptides by hydrolytic enzymes also takes place. Consequently, the vacuole plays a key role in both cellular metabolism and homeostasis of the cytosolic pH and ion balance (8, 9).

Transport Activities Across Biological Membranes

Storage, import and export of metabolites and ions, respectively, require corresponding transport systems across biological membranes. Balance between net cellular and vacuolar accumulation and release for a given ion will be determined by the relative activities of three classes of transport systems. The simplest case is the transport of the given ion which passively follows thermodynamically downhill its concentration gradient via ion-selective channel proteins in membrane. Ion channels are water-filled pores usually with a discrete gating behaviour and are controlled by ligands, voltage or mechanical pressure. Carriers couple the uphill flux of ions to that of protons which follow their opposite directed downhill concentration out of the vacuole. The third class is that of pumps which couple the energy from ATP hydrolysis to the uphill transport of protons into the vacuole (10).

A small number of substrate transporting proteins are known in the vacuolar membrane (Fig. 1). In the yeast membrane transporter's community, a Ca^{2+} dependent and Ca^{2+} permeable cation channel, called YVC1, is probably the best known ion channel of the tonoplast. This channel has been extensive studied and described by patch clamp experiments (11, 12, 13, 14). The accumulation of amino acids and calcium in the vacuolar lumen has been demonstrated by biochemical means and corresponding transport systems are assumed to be proton-substrate antiporters for which the encoding gene for the H^+/Ca^{2+} exchanger (VCX1) has been found. There is to date are no genetic evidence for the amino acid antiporters (15, 16, 17, 18). An alternative Ca^{2+} uptake system is postulated from genetic studies which revealed an ATP dependent Ca^{2+} pump (PMC1) (19). Another ATP dependent transport system, a Glutathion-S-conjugate transporter YCF1, has been also postulated from genetic studies, and sequence alignments with known genomic sequences showed that this transporter belongs to the great family of ABC transporters (18). Besides NHX1, a H^+/Na^+ antiporter, which has only been postulated from biochemical transport studies without any genetic evidence, a further series of transporters for phosphate, sulphate and chloride have been postulated for the proper functioning of the entire vacuole but their existence has not been proven either from transport studies or from genetic studies.

The transport of substrates through the antiporters mentioned above is coupled to an electrochemical proton gradient across the tonoplast which is generated by a vacuolar proton-translocating adenosine triphosphate hydrolase (V-type H+-ATPase, TC 3.2.2) (18, 20). The H^+-ATPase couples the energy from ATP hydrolysis to the uphill transport of protons from the cytosol into the vacuole.

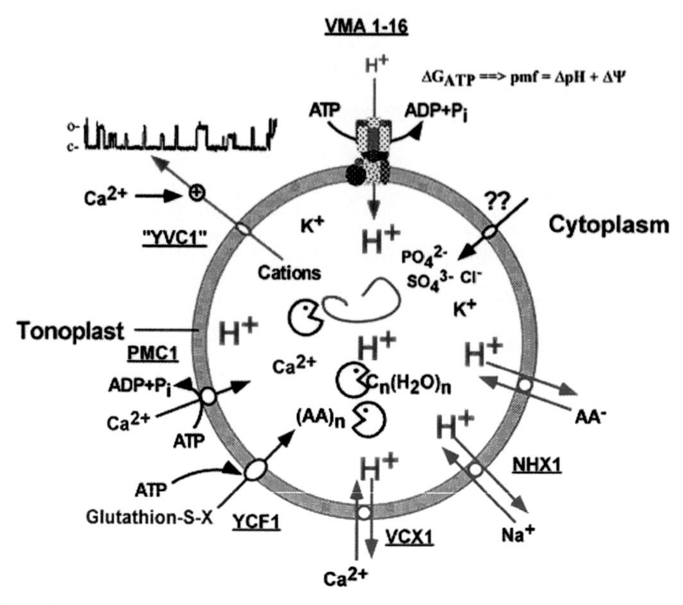

Figure 1. Schematic representation of the yeast vacuole with tonoplast-residing transport systems.

This proton translocation results in acidification of the vacuolar lumen and thus to the generation of a pH gradient across the membrane. Furthermore, the accumulation of positive charges within the vacuole creates an electrical potential difference across the membrane (which is defined as voltage) with positive voltage inside the vacuole. Voltage together with pH gradient build up a driving force, called proton motive force (*pmf*) (21), which in the case of the tonoplast results in low vacuolar pH creating a cytoplasm-directed proton gradient rather than in the generation of an electric voltage across the membrane due to parallel ion conducting transport systems with equilibrium voltages more positive than the maximum ATPase generated voltage (22).

MOLECULAR STRUCTURE OF THE V-ATPASE

Extensive studies on the structure and function of the pump have been carried out in detail by optical, biochemical and genetic methods on *S. cerevisiae*, *Neurospora crassa* (a fungus), as well as on plant and animal cells (for reviews, see: 6, 23, 24, 25, 26). Electron microscopic images of the ATPase holoenzyme show that the pump consists of at least two distinct sectors of which a peripheral domain is readily distinguishable from the integral membrane domain. The peripheral domain is connected by a stalk structure to the membrane-bound domain. Cross-sections of fungal vesicles showed that these ball-and-stalk structures seemed to be arranged like a string of pearls on the surface of the membrane (25, 27). From optical investigations it could be demonstrated that, under certain conditions, the holoenzyme decays into its two main sectors (28).

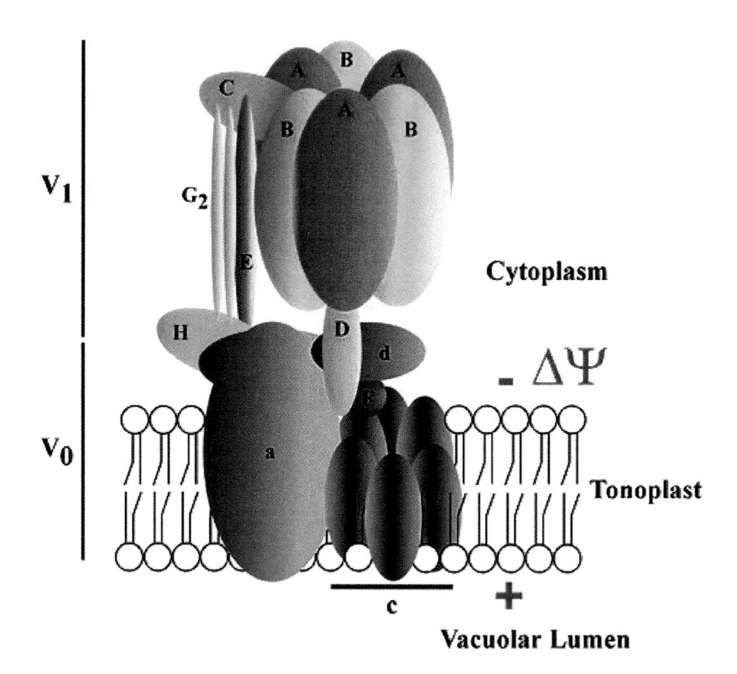

Figure 2. Topological model of the yeast V-ATPase (according to Arata *et al.* (53)). Subunits in the peripheral domain V_1 are indicated by capital letters, whereas subunits in the integral domain V_O are indicated by small letters.

Biochemical and genetic experiments revealed that the pump is a multisubunit enzyme complex composed of at least 7 different subunits (peripheral domain, called V_1) with stoichiometry A_3B_3CDEFG and at least 4 different subunits (integral part, called V_O), respectively (Fig. 2). The subunits A (69 kDa) and B (59 kDa) (encoded by the genes VMA1 and VMA2 (29, 30)) are the largest subunits of the V_1-domain which is the catalytic domain of the pump. Here, ATP binding and hydrolysis takes place (31). At the V_O domain a remarkable structural feature should be noted: the subunit c (VMA3, 17 kDa) forms – six-fold copied – a barrel shaped hexamer which forms the proton channel (32). The molecular weight of the pump is between 500 and 750 kDa, and the yeast V-ATPase is encoded by at least 16 different genes (33). ATP-dependent proton translocation by the pump was first observed by biochemical transport studies carried out with voltage- and pH-sensitive dyes (34).

However, the key determinant of the pump, the transport coupling ratio, is best estimated by electrophysiological methods. Detailed studies were first reported from plant V-ATPases because plant vacuoles with diameters up to 100 µm are more suited for patch clamp experiments than yeast vacuoles (10 to 15 µm in diameter), thus, little is known about the biophysical properties of the yeast pump.

In this overview, I shall expand on our preliminary report (34) and present some details of the yeast V-ATPase obtained from patch clamp experiments with special emphasis on the transport coupling ratio (Kettner & Bertl, submitted).

PATCH CLAMP EXPERIMENTS AND CELL PREPARATION

A common property of all the transport systems mentioned previously is that they transport charged substrates. Transport of these substrates, mostly ions, is recordable as an electrical current whose size is dependent on the basic electrical quantities such as membrane voltage and resistance of the conducting transport systems. The patch clamp technique allows recording of currents from a small membrane area (patch) in response to a defined command voltage (voltage clamp).

The investigation of the yeast ATP-driven proton transport by the V-ATPase has been carried out with the patch clamp technique in the whole-vacuole configuration (35).

Vacuoles were released from protoplasts in the recording chamber by subjecting 24-hour-old protoplasts to mild acid lysis in low-Ca^{2+} medium. Shortly after the first vacuoles have been released, releasing solution was replaced by standard recording buffer (150 mM KCl, 5 mM $MgCl_2$, 0.5 mM EGTA, pH 7 with Tris/MES) which removes residual cell fragments, lipids and dirt. A standard solution-filled glass microelectrode, the patch pipette, was connected via the pipette holder with a pre-amplifier and the movement of this recording system was controlled by micromanipulator (Fig. 3). The patch pipette was firmly attached to the tonoplast which resulted in a high mechanical stability and electrical resistance of up to 5 gigaohms (1 $G\Omega = 10^9\ \Omega$), called giga-seal. A tight seal is required for the current recording with high signal/noise ratio to make sure that the small transport currents are masked neither by statistical noise nor by leakage currents between pipette and membrane.

This seal formation is called cell-attached configuration and determines the starting point for further configurations (36, 37). Establishment of the whole-vacuole configuration required that the membrane underlying the pipette tip was ruptured by a brief voltage pulse (600 mV, 3 ms) whereas the seal resistance has to be kept significantly above 1 GΩ. The breakage was monitored by the change of the electrical properties of the glass-membrane system which are determined by the appearance of "slow" capacitative transients whose size is dominated by charge movement within the vacuolar membrane (35). The whole vacuole configuration allows current recording from the entire vacuolar membrane with defined voltages.

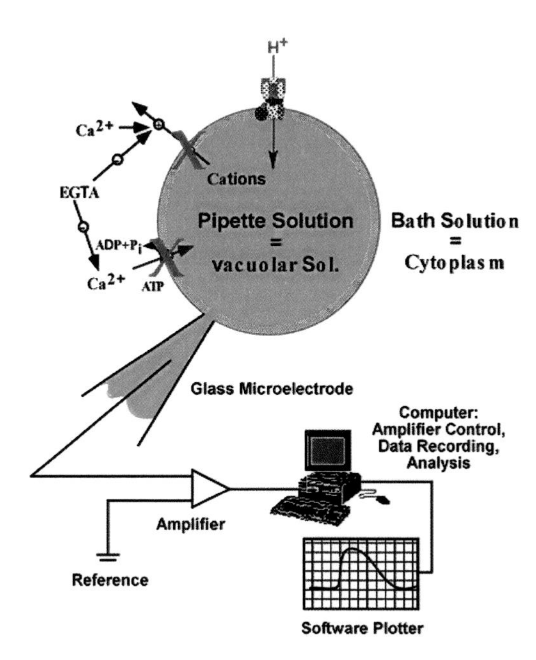

Figure 3. Representation of the experimental setup with patch clamp whole-vacuole configuration. EGTA in bath and pipette solution avoids activation by calcium of both the cation channel and the Ca^{2+} pump and thus masking pump currents.

Due to the larger volume of the patch pipette compared to that of the vacuole, the vacuolar solution was completely exchanged by the pipette solution within a few seconds so that the composition of the vacuolar solution can be assumed to be known (38). The cytosolic side of the tonoplast is exposed to the bath solution which simulates the cytosol and matches the vacuolar solution (standard recording solution, see above). This symmetric composition avoids the build up of any driving forces for ion transport across the tonoplast. Furthermore, the calcium concentration in pipette and bath solution was held at virtually zero by EGTA to avoid masking the smaller pump currents by the much larger channel currents from the cation channel YVC1. For continous data recording a software package for Macintosh (HEKA, Pulse/PulseFit

8.0) was used, supplemented with the chart recorder extension X-Chart, in combination with the EPC-9/ITC-16 amplifier/data acquisition system. Data were filtered at 100 Hz with a built-in 8-pole Bessel filter, sampled at 1 kHz and stored on the computer hard drive.

Determination of the transport coupling ratio was carried out by analysis of current-voltage (I/V) characteristics of the vacuolar membrane under different cytosolic and vacuolar pH conditions. Currents were recorded in response to a voltage ramp which clamped the membrane potential gradually from –80 to +80 mV within 2.5 s and were plotted against the applied voltage. The I/V characteristics were obtained both in the absence and presence of ATP/ADP and P_i when the ATP-dependent currents reached their maximum. Subtraction of both characteristics yields the I/V characteristics of the vacuolar H^+-ATPase. The sign convention for membrane voltage and current, as proposed by Bertl *et al.* (39) was used throughout. This convention defines a positive current (= outward current) as the flow of positive charges from the cytoplasmic side of a membrane to the extra-cytoplasmic side, which can be both the extra-cellular area and intra-organellar volume. This positive current is drawn upwards in all representations of current traces and I/V plots.

A tetraploid strain of *Saccharomyces cerevisiae* (YCC78, *MAT*a,a,α,α, *ura3-52 ade2-101*, (40)) was used throughout since it contains larger cells than haploid strains. It should be emphasized however, that patch clamp experiments are usually not limited by the size of the cells. Nevertheless, large cells are more convenient and easier to work with than small cells. General methods for growing, handling and protoplasting yeast and for isolating vacuoles have previously been described in detail (34).

Results and Discussion

The whole vacuole configuration was obtained in symmetric standard solutions (150 mM KCl, 5 mM $MgCl_2$, 0.5 mM EGTA, pH 7 with Tris/MES). The membrane voltage was clamped at 0 mV and current recordings were commenced when the membrane current was stable at approximately 0 pA. Addition of 5 mM ATP to the bath solution induced a nearly instantanous current after a short lag time of about 10 to 15 s (Fig. 4). This *per definitionem* outward current reached a maximum of up to 15 pA ± 6 pA (n=10 vacuoles), corresponding to about 30 mAm^{-2} ± 7 mAm^{-2}. These current values are comparable to the plant V-ATPase current densities which are reported to be between 5 mAm^{-2} and 23 mAm^{-2} (41, 42).

It is remarkable that the current declined slowly and reached the baseline level within 15 to 20 min in the sustained presence of ATP. There is evidence that the inhibition of the pump results from tightly bound ADP at the catalytic side of the enzyme rather than by a chaotropic effect which has been reported for plant V-ATPases (54). For example, high salt concentrations – namely with chaotropically acting dissociated anions such as Cl^- and NO_3^- - in presence of ATP, have been shown to be responsible for the dissociation of the V_1 and V_O subunits of plant, fungal and animal V-ATPases (27, 43, 44). The loss of the activity of the V-ATPase and the regain, is reported to be a controlled mechanism in response to changed extracellular conditions.

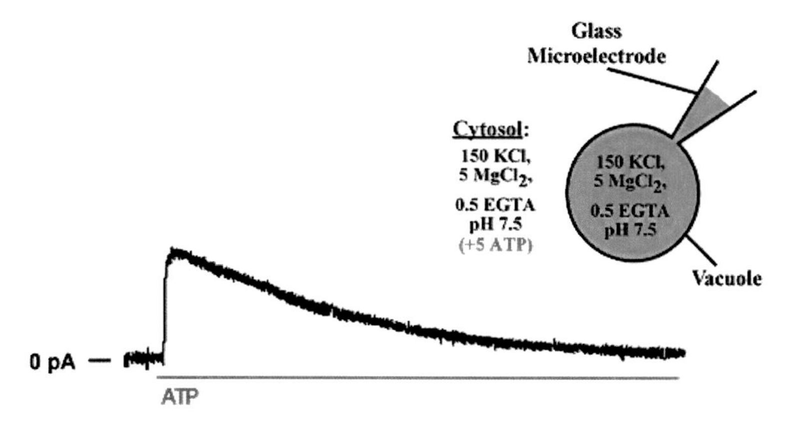

Figure 4. Whole vacuolar recording of ATP dependent current at 0 mV clamping voltage. The current reaches a maximum and decreases with ATP continuously present (bar of the bottom of the current trace).

In order to see whether this ATP-induced current was indeed generated by the V-type ATPase and not by other ATP-dependent transport systems inserted within the tonoplast, experiments were conducted in the presence of bafilomycin A_1. This substance and other bafilomycin-derived agents are known to be highly specific and potent inhibitors of V-type ATPases, whilst F-type and P-type ATPases are only slightly or not at all affected (45, 46). The addition of 100 nM bafilomycin A_1 to the bath solution containing ATP at the peak of the ATP-induced current resulted in a rapid decline of this current to the zero current level (Fig. 5). This complete inhibition of the ATP-induced current is evidence for the supposition that this current was indeed solely generated by the V-ATPase. Consequently, bafilomycin is a well-suited candidate to distinguish between V-ATPase activity and the activity of other ATP-dependent transport systems. Under these experimental conditions, the activity of other ATP-dependent transport system, such as Ca^{2+}-ATPase (PMC1) or ABC-transporter YCF1, was not detected.

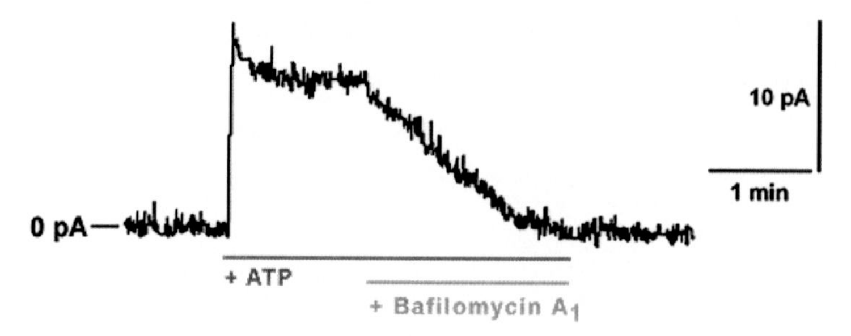

Figure 5. The ATP-induced whole vacuolar currents are bafilomycin-sensitive indicating that the ATP driven current is mediated by the vacuolar H^+-ATPase.

For definite identification of the ATP dependent current as the activity of the V-ATPase, it was necessary to study the key thermodynamic determinant of a pump which is the transport coupling ratio. The transport coupling ratio, which is often incorrectly described as stoichiometry of transports, is defined as the number of protons per ATP molecule hydrolysed.

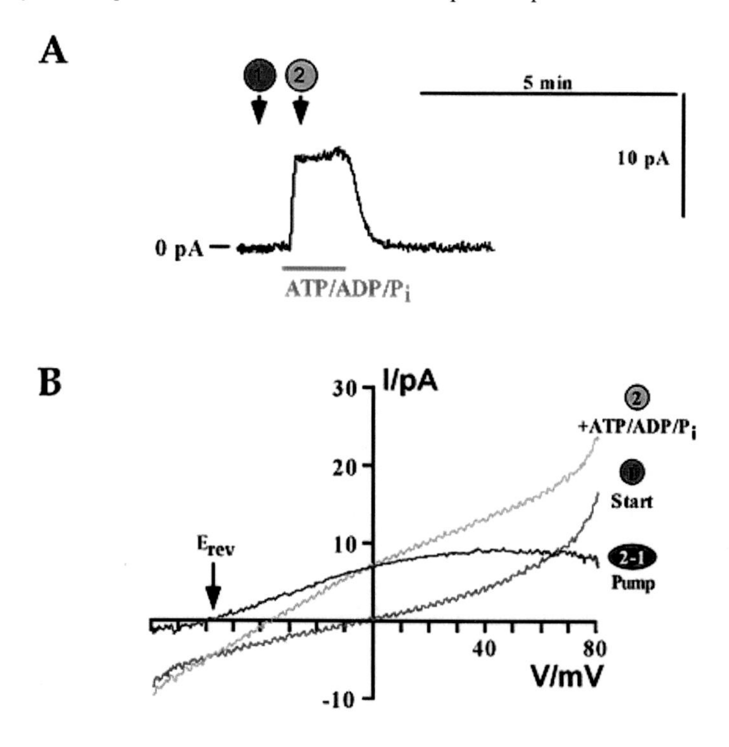

Figure 6. Determination of E_{rev} for the V-ATPase in presence of ATP, ADP and P_i (I). **A)** Whole vacuolar current trace at 0 mV in standard solution at pH_{cyt} 7.5 and pH_{vac} 5, and subsequent additon of 5 mM ATP, 5 mM ADP and 10 mM P_i. **B)** Current-voltage (I/V) characteristics of the vacuolar membrane in absence (*1*) and presence (*2*) of ATP/ADP and P_i. Subtraction (*2-1*) yields the pump characteristics. The intersection of this I/V curve with the voltage axis determines E_{rev}.

For example, H^+-translocating P-type ATPases in the plasma membrane of plants and fungi energize the transport of a single proton by the hydrolysis of one molecule ATP. They generate a membrane voltage up to –400 mV which is used to drive other transport systems, however, the pH gradient yields only about one pH unit. By contrast, plant V-ATPases transport 2 to 3 protons under hydrolysis of one molecule ATP and generate a steep pH gradient of up to 5 pH units (as for example in the vacuoles of lemon fruits (47)) but the voltage across the tonoplast yields only to –30 to –50 mV (see also Section 4).

According to previously published methods (48), the estimation of the coupling ratio was carried out by analysis of the I/V characteristics of the vacuolar membrane under different cytosolic and vacuolar pH conditions. The characteristics were obtained both in absence of ATP/ADP and P_i and in the presence of 5 mM ATP, 5 mM ADP and 10 mM P_i after the current reached its maximum (Fig. 6A).

As depicted in Fig. 6B, the first curve of the I/V plot shows the electrical properties of the entire vacuolar membrane which are determined by the ATP independent transport systems. The second trace consists of the sum of ATP dependent and independent transport systems. The result from the subtraction of both these current traces gives the characteristics of the pump which are marked by (i) sigmoidal shape with saturation of the current towards the limit of the applied voltage, (ii) the corresponding short circuit current at 0 mV in the current trace (Fig. 6A), and (iii) the intersection of the I/V curve with the voltage axis which represents the reversal potential of the pump and determines the thermodynamic equilibrium at which no ATP-dependent proton net-flux occurs. The value of the reversal potential was then used to calculate the coupling ratio according following equation:

$$\Delta\Psi = E_{rev} = \frac{59mV}{n} * \left(\log\left(K_{ATP} * \frac{[ATP]}{[ADP]*[P_i]} \right) + n * \log\left(\frac{[H^+]_c}{[H^+]_v} \right) \right)$$

where $\Delta\Psi$ is the membrane potential and corresponds to the reversal potential (E_{rev}), n is the coupling ratio of H^+ translocated per ATP hydrolysed, K_{ATP} is the equilibrium constant for ATP hydrolysis, the square brackets denote the activities of ATP, ADP and inorganic P (P_i) and the subscripts c and v refer to the cytosolic and vacuolar solution, respectively. The value for K_{ATP} depends strongly on pH and free pMg^{2+} of the ATP solution and was calculated using ΔG^0, the values for the free energy of ATP hydrolysis, given by the following equations:

$$\Delta G^0 = -RT * \ln K_{ATP}$$
$$\Rightarrow K_{ATP} = e^{-\Delta G^0 / RT}.$$

The values for ΔG^0 were obtained from Alberty (49) and, when pH and pMg^{2+} were taken into account, yielded the corresponding values for K_{ATP}.

Fig. 7 shows a set of several I/V characteristics of the pump at different cytosolic pH values and at a constant vacuolar pH of 5. The reversal potential of the pump is shifted toward negative voltages at higher pH values of the cytosol, i.e. from –20 mV (pa_{ct} 8.5) to –60 mV (pa_{ct} 6). At symmetrical pH 5, E_{rev} could not be determined because the intersection of the current curve with the voltage axis might be outside of the applied voltage. Table 1 shows that corresponding to the shift of the reversal potential, an increase of the calculated coupling ratios occurs from 2.5 H^+/ATP at pH 8.5 to 4.1 H^+/ATP at pH 6.

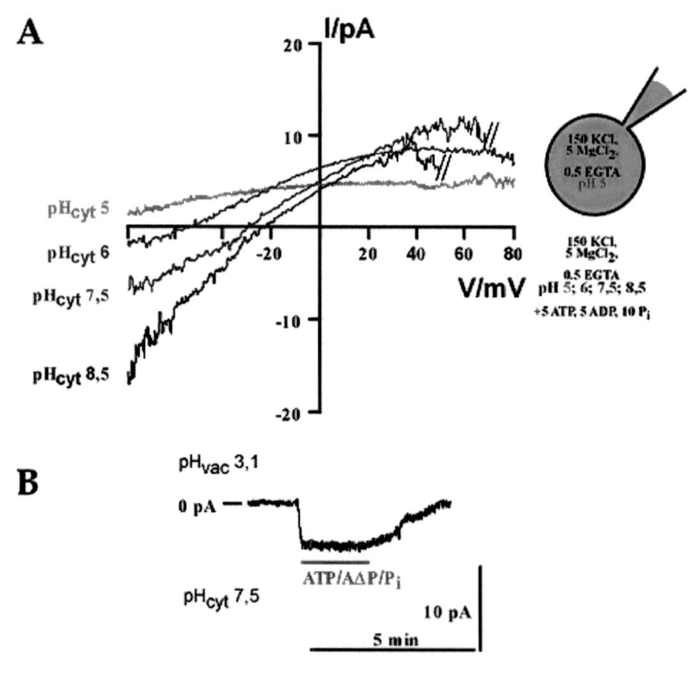

Figure 7. Determination of E_{rev} of the V-ATPase (II). **A)** Set of I/V characteristics of the pump showing dependence on cytosolic pH at constant vacuolar pH 5. E_{rev} shifts towards negative voltages with acidification of the cytosol. These E_{rev} values were used to calculate the transport coupling ratios. **B)** With the membrane voltage held at 0 mV and with a vacuolar pH of 3.1, the addition of ATP/ADP and P_i induced an inward current. The pump worked obviously in the reverse mode suggesting ATP synthesis.

At various vacuolar pH and at constant cytosolic pH of pH 7.5, the analysis of the pump characteristics shows a shift of the reversal potential towards positive voltages with acidification of the vacuolar lumen, i.e. from –40 mV (pH_{vac} 6) to –25 mV (pH_{vac} 4.1).

At symmetrical pH 7.5, the intersection of the current curve could also not be determined. Corresponding to the shift of the reversal potentials, the calculated coupling ratios decreased from 4.1 H^+/ATP at pH 6 to 2.3 H^+/ATP at pH 4.1 (Tab. 1).

It is remarkable that with the membrane voltage held at 0 mV and with a vacuolar pH of 3.1, the addition of ATP induced an inward current, out of the vacuole into the cytoplasm (Fig. 7). The pump worked obviously in the reverse mode under these conditions which suggests ATP synthesis coupled to the translocation of about 2.5 protons, however, we did not carry out any experiments to detect synthesized ATP molecules. This reversible behaviour of the pump has also been reported from plant V-ATPases (50).

Under physiological conditions (pH_{cyt} 7 to 8 and pH_{vac} 4 to 5), the coupling ratios were estimated to be between 2 and 3 H^+/ATP.

Table 1. Coupling ratios of the V-ATPase in dependence of ΔpH across the vacuolar membrane

pH_{vac} 5	**pH_{cyt}**	**E_{rev} ± SE (mV)**	**Coupling Ratios ± SE H^+/ATP**
	8.5	-20.2 ± 4	2.55 ± 0.05
	7.5	-27.7 ± 5.1	2.98 ± 0.1
	6	-58.3 ± 9.1	4.15 ± 0.3
	5	n.d.	n.d.
pH_{cyt} 7.5	**pH_{vac}**	**E_{rev} ± SE (mV)**	**Coupling Ratios ± SE H^+/ATP**
	7.5	n.d.	n.d.
	6	-37.3 ± 5.5	4.15 ± 0.2
	5	-31.8 ± 7.8	2.91 ± 0.12
	4.1	-25.7 ± 4.1	2.31 ± 0.05
	3.1	+50 ± 8	2.49 ± 0.09

The data show that the V-ATPase incompletely couples proton transport across the vacuolar membrane to the hydrolysis of ATP. The coupling ratios are non-integer, variable and dependent on ΔpH across the vacuolar membrane. These properties of the coupling ratios as well as their values are consistent with those found in plant vacuoles (41, 51). This phenomenon can be best described as slippage and tunnelling of the pump. The slippage effect was observed for both plant and yeast V-ATPase: Due to the reduced acidification of the cytoplasm, the decreased coupling ratio demonstrates that the pump remains catalytically active even although occupancy of the H^+ binding sites was incomplete. By contrast, tunnelling, which is described by increasing coupling ratios when the cytoplasm was acidified, indicates that protons may have overcome the tonoplast without coupling to ATP hydrolysis (52).

The yeast V-ATPase shows high coupling ratios and low membrane voltages at pump equilibrium. This indicates that the capacity of the pump generates a steep proton gradient rather than high membrane voltages. Thus, the V-ATPase effectively maintains cytosolic pH homeostasis and generates a powerful *pmf* to drive the co transport of other substrates across the tonoplast.

CONCLUSION

The results presented here, demonstrate the feasibility of the electrophysiological recording techniques developed to investigate both the biophysical properties of ATP-driven active transport across the tonoplast of *S. cerevisiae* and other transport systems lacking channel properties such as cotransporters with a coupling ratio greater than 1.

Furthermore, this work also shows that computers are powerful tools for purposes other than doing *in-silico* biology, retrieving databases and modelling.

In the area of electrophysiology computers and appropriate software packages effectively help doing experiments by recording data in nearly real-time. Online analysis during the current experiments allows redesigning of the experimental setup if required. Collecting experimental data by computers allows storage and archiving of huge amounts of data on hard disks. Dependent of the sampling rate and filter setting as well as the extent of the experimental protocol, data amounts of up to 5 MB per experiment have to be stored. The great advantage of digitalisation of experimental data is the simple access to these data for analysis and preparation for publication.

Furthermore, the integration of hardware functions such as filter, oscilloscope, and voltage generator into software greatly reduces the equipment setup. Last but not least, this configuration makes it easy to combine electrophysiological approaches simultaneously with visualisation methods such as Ca^{2+}-imaging.

In conclusion, the predominant aim of this review was to show that only a combination of diverse methods and techniques as well tools – *in-silico* and "in-reality" – yield a comprehensive insight to the structure and function of complex molecules such as the vacuolar proton pump.

ACKNOWLEDGMENTS

I thank Adam Bertl (University of Karlsruhe, Germany) for generously providing his lab for my patch clamp experiments and Allan Dunn (Beilstein CD&S, Frankfurt/Main) for helpful discussions. The experimental part of this work has been supported by research grant Be1181/ 4-1 from the Deutsche Forschungsgemeinschaft (DFG).

REFERENCES

[1] Bradshaw, R. A. (2002). Proteomics – Boom or bust? *Molecular & Cellular Proteomics* **1**(3):177-178.

[2] Nelissen, B., Mordant, P., Jonniaux, J.-L., De Wachter, R., Goffeau, A. (1995). *FEBS Lett.* **377**:232-236.

[3] André, B. (1995). An overview of membrane transport proteins in *Saccharomyces cerevisiae*. *Yeast* **11**:1575 – 1611.

[4] Bassett, D. E., Jr. et al. (1997). Genome cross-referencing and XREFdb : Implications for the identification and analysis of genes mutated in human disease. *Nature Genetics* **15**:339-344.

[5] Klionsky, D. J., Nelson, H., Nelson, N. (1992). Compartment acidification is required for efficient sorting of proteins to the vacuole of *Saccharomyces cerevisiae*. *J. biol. Chem.* **267**:3416-3422.

[6] Klionsky, D. J. , Herman, P. K., Emr. S. D. (1990). The fungal vacuole: Composition, function and biogenesis. *Microbiol. Rev.* **54**:266-292.

[7] Bowman, B. J., Vazquez-Laslop, N., Bowman, E.J. (1992). The vacuolar ATPase of *Neurospora crassa*. *J. Bioenerg. Biomembr.* **24**:361-369.

[8] Davies, R. H. (1986). Compartmental and regulatory mechanisms in the arginine pathways of *N. crassa* and *S. cerevisiae*. *Microbiol. Rev.* **50**:280-313.

[9] Anraku, Y., Umemoto, N. Hirata, R., Wada, Y. (1989). Structure and function of the yeast vacuolar membrane proton ATPase. *J. Bioenerg. Biomembr.* **21**:589-603.

[10] Hille, B. (1992). Ion channels of excitable membranes. Sinauer Ass. Inc. Sunderland.

[11] Bertl, A. & Slayman, C. L. (1990). Cation-selective channels in the vacuolar membrane of *Saccharomyces*: Dependence on calcium, redox-state and voltage. *Proc. Natl. Acad. Sci. USA* **87**:7824-7828.

[12] Bertl, A., Gradmann, D., Slayman, C. L. (1992). Calcium- and voltage-dependent ion channels in *Saccharomyces cerevisiae*. *Phil. Trans. R. Soc. Lond.* **338** :63-72.

[13] Wada, Y., Ohsumi, Y., Tanifuji, M., Kasai, M., Anraku, Y. (1987). Vacuolar ion channel of the yeast, *Saccharomyces cerevisiae. J. biol. Chem.* **262**:17260-17263.

[14] Tanifuji, M., Sato, M., Wada, Y., Anraku, Y., Kasai, M. (1988). Gating behaviours of a voltage-dependent and Ca^{2+}-activated cation channel of yeast vacuolar membrane incorporated into planar lipid bilayer. *J. Membr. Biol.* **106**:47-55.

[15] Ohsumi, Y. & Anraku, Y. (1981). Active transport of basic amino acids driven by a proton motive force in vacuolar membrane vesicles of *S. cerevisiae. J. biol. Chem.* **258**:2079-2082.

[16] Ohsumi, Y. & Anraku, Y. (1983). Calcium transport driven by a proton motive force in vacuolar membrane vesicles of *Saccharomyces cerevisiae. J. biol. Chem.* **258**:5614-5617.

[17] Sato, T., Ohsumi, Y., Anraku, Y. (1984). An arginine/histidine exchange transport system in vacuolar membrane vesicles of *Saccharomyces cerevisiae. J. biol. Chem.* **259**:11509-11511.

[18] Sato, T., Ohsumi, Y., Anraku, Y. (1984). Substrate specificities of active transport systems for amino acids in vacuolar-membrane vesicles of *Saccharomyces cerevisiae. J. biol. Chem.* **259**:11505-11508.

[19] Paulsen, I. T., Sliwinski, M. K., Nelissen, B., Goffeau, A., Saier, M. H. Jr. (1998). Unified inventory of established and putative transporters encoded within the complete genome of *Saccharomyces cerevisiae. FEBS Lett.* **430**:116-125.

[20] Kakinuma, Y., Ohsumi, Y., Anraku, Y. (1981). Properties of H^+-translocating adenosine triphosphatase in vacuolar membrane of *Saccharomyces cerevisiae. J. biol. Chem.* **256**:10859-10863.

[21] Mitchell, P. & Moyle, J. (1965). Stoichiometry of proton translocation through the respiratory chain and adenosine triphosphate systems of rat liver mitochondria. *Nature* **208**:147-151.

[22] Nelson, N., Perzov, N., Cohen, A., Hagai, K., Padler, V., Nelson, H. (2000). The cellular biology of proton-motive force generation by V-ATPases. *J. exp. Biol.***203**:89-95.

[23] Nelson, N. & Harvey, W. R. (1999). Vacuolar and plasma membrane proton-adenosinetriphosphatases. *Physiological Reviews* **79**(2): 361-385.

[24] Sze, H., Ward, J. M., Lai, S. (1992). Vacuolar H^+-translocating ATPases from plants: Structure, function and isoforms. *J. Bioenerg. Biomembranes* **24**(4): 371-380.

[25] Stevens, T. H. & Forgac, M. (1997). Structure, function and regulation of the vacuolar H^+-ATPase. *Annu. Rev. Cell Dev. Biol.* **13**:779-808.

[26] Bowman, B. J., Dschida, W. J., Bowman, E. J. (1992). Vacuolar ATPase of *Neurospora crassa*: Electron microscopy, gene characterization and gene inactivation/mutation. *J. exp. Biol.* **172**:57-66.

[27] Lüttge, U., Fischer-Schliebs, E., Ratajczak, R., Kramer, D., Berndt, E., Kluge, M. (1995). Functioning of the tonoplast vacuolar C-storage and remobilization in crassulacean acid metabolism. *J. exp. Bot.* **46**:1377-1388.

[28] Nelson, N. (1992). The vacuolar H^+-ATPase – one of the most fundamental ion pumps in nature. *J. exp. Biol.* **172**:19-27.

[29] Hirata, R., Ohsumi, Y., Nakano, A., Kawasaki, H., Suzuki, K., Anraku, Y. (1990). Molecular structure of a gene, VMA1, encoding the catalytic subunit of H^+-translocating adenosine triphosphatase from vacuolar membranes of *Saccharomyces cerevisiae. J. biol. Chem.* **265**:6726-6733.

[30] Kane, P. M., Yamashiro, C. T., Wolczyk, D. R., Neff, N., Goebl, M., Stevens, T. H. (1990). Protein splicing converts the yeast TFP1 gene product to the 69-kDa subunit of the vacuolar H^+-adenosine triphosphatase. *Science* **250**:651-657.

[31] Webster, L. C. & Apps, D. K. (1996). Analysis of nucleotide binding by a vacuolar proton-translocating adenosine triphosphatase. *Eur. J. Biochem.* **240**:156-164.

[32] Nelson, N. (1989). Structure, molecular genetics and evolution of vacuolar H^+-ATPase. *J. Bioenerg. Biomembr.* **21**:553-571.

[33] Mewes, H. W., Frishman, D., Güldener, U., Mannhaupt, G., Mayer, K., Mokrejs, M., Morgenstern, B., Münsterkoetter, M., Rudd, S., Weil, B. (2002). MIPS: a database for genomes and protein sequences. *Nucleic Acids Research* **30**(1):31-4; http://mips.gsf.de/proj/yeast/catalogues/funcat/fc40_25.html.

[34] Uchida, E., Ohsumi, Y., Anraku, Y. (1985). Purification and properties of H^+-translocating, Mg^{2+}-adenosine triphosphatase from vacuolar membranes of *Saccharomyces cerevisiae. J. biol. Chem.* **260**:1090-1095.

[35] Bertl, A., Bihler, H., Kettner, C., Slayman, C. L. (1998). Electrophysiology in the eukaryotic model cell *Saccharomyces cerevisiae. Pflügers Arch.* **436**:999-1013.

[36] Hamill, O. P., Marty, A., Neher, E., Sakmann, B., Sigworth, F. J. (1981). Improved patch clamp techniques for high resolution current recording from cells and cell-free membrane patches. *Pflügers Arch.* **391**:85-100.

[37] Sakmann, B. & Neher, E. (1995). Single channel recording. 2nd Ed. Plenum, New York.

[38] Marty, A. & Neher, E. (1983). Tight seal whole-cell recording. Chapter 7 in: Single Channel Recording. 1st Ed. Editors Sakman, B. & Neher, E., Plenum Press New York.

[39] Bertl, A., Blumwald, E., Coronado, R., Eisenberg, R., Findlay, G., Gradmann, D., Hille, B., Köhler, K., Kolb, H.-A., MacRobbie, E., Meissner, G:, Miller, C., Neher, E., Palade, P., Pantoia, O., Sanders, D., Schroeder, J., Slayman, C. L., Spanswick, R., Walker, A., Williams, A. (1992). Electrical measurements on endomembranes. *Science* **258**:873-874.

[40] Mirzayan, C., Copeland, C. S., Snyder, M. (1992). The NUF1 gene encodes an essential coiled-coil related protein that is a potential component of the yeast nucleoskeleton. *J. Cell Biol.* **16**:1319-1332.

[41] Coyaud, L., Kurkdjian, A., Kado, R., Hedrich, R. (1987). Ion channels and ATP-driven pumps involved in ion transport across the tonoplast of sugarbeet vacuoles. *Biochem. Biophys. Acta* **902**:263-268.

[42] Davies, J. M., Hunt, I., Sanders, D. (1994). Vacuolar H^+-pumping ATPase variable transport coupling ratio controlled by pH. *Proc., Natl. Acad. Sci. USA* **91**:8547-8551.

[43] Puoplo, K. & Forgac, M. (1990). Functional reassembly of the coated vesicle proton pump. *J. Biol. Chem.* **265**:14836-14841.

[44] Kane, P. M. (2000). Regulation of V-ATPases by reversible disassembly. *FEBS Lett.* **469**:137-141.

[45] Bowman, E. J., Siebers, A., Altendorf, K. H. (1988). Bafilomycins: A class of inhibitors of membrane ATPases from microorganisms, animal cells and plant cell. *Proc. Natl. Acad. Sci. USA* **85**:7972-7975.

[46] Dröse, S., Bindseil, K. U., Bowman, E. J., Siebers, A., Zeeck, A., Altendorf, K. (1993). Inhibitory effect of modified bafilomycins and concanamycins on P- and V-type ATPases. *Biochemistry* **32**:3902-3906.

[47] Sinclair, W. B. (1984). The biochemistry and physiology of the lemon and other citrus fruits, pp. 109 – 156, University of California, Division of Agriculture and Natural Resources, Oakland, CA.

[48] Rea, P. A. & Sanders, D. (1987). Tonoplast energization: two H^+ pumps, one membrane. *Physiologia Plantarum* **71**:131-141.

[49] Alberty R. A. (1968). Effect of pH and metal ion concentration on the equilibrium hydolysis of adenosine triphosphate to adenosine diphosphate. *J. biol. Chem.* **243**:1337-1343.

[50] Hirata, T., Nakamura, N., Omote, H. Wada, Y., Futai, M. (2000). Regulation and reversibility of vacuolar H^+-ATPase. *J. biol. Chem.* **275**:386-389.

[51] Müller, M. L., Jensen, M., Taiz, L. (1999). The vacuolar H^+-ATPase of lemon fruits is regulated by variable H^+/ATP coupling and slip. *J. biol. Chem.* **274**:10706-10716.

[52] Läuger, P. (1991). Electrogenic ion pumps. Sinauer Ass. Inc. Sunderland.

[53] Arata, Y., Baleja, J. D., Forgac, M. (2002). Cysteine-directed cross-linking to subunit B suggests that subunit E forms part of the peripheral stalk of the vacuolar H^+-ATPase. *J. biol. Chem.* **277**:3357-3363.

[54] Kettner, C., Obermeyer, G., Bertl, A. (2003). Inhibition of the yeast V-type ATPase by cytosolic ADP. *FEBS Lett.* **535**:119 – 124.

PROTEIN MISFOLDING AND ITS LINKS WITH HUMAN DISEASE

CHRISTOPHER M. DOBSON

Department of Chemistry, University of Cambridge, Lensfield Road, Cambridge CB2 1EW, UK
E-Mail: cmd44@cam.ac.uk

Received: 18th June 2002 / Published: 15th May 2003

ABSTRACT

The ability of proteins to fold to their functional states following synthesis on the ribosome is one of the most remarkable features of biology. The sequences of natural proteins have emerged through evolutionary processes such that their unique native states can be found very efficiently even in the complex environment inside a living cell. But under some conditions proteins fail to fold correctly, or to remain correctly folded, in living systems, and this failure can result in a wide range of diseases. One group of diseases, known as amyloidoses, which includes Alzheimer's and the transmissible spongiform encephalopathies, involves deposition of aggregated proteins in a variety of tissues. These diseases are particularly intriguing because evidence is accumulating that the formation of the highly organized amyloid aggregates is a generic property of polypeptides, and not simply a feature of the few proteins associated with recognized pathological conditions. Moreover, such aggregates appear to posses inherent toxicity. That aggregates of this type are not normally found in properly functional biological systems is a further testament to the efficiency of biological evolution, in this case resulting in the emergence of a variety of mechanisms inhibiting their formation. Understanding the nature of such protective mechanisms is a crucial step in the development of strategies to prevent and treat these debilitating diseases.

PROTEIN FOLDING AND MISFOLDING

A living organism may contain as many as 50,000 different types of protein. Following synthesis on the ribosome, each protein molecule must fold into the specific conformational state that is encoded in its sequence in order to be able to carry out its biological function. How this process happens is one of the most fascinating and challenging problems in structural biology (1, 2). In the cell, folding takes place in a complex and highly crowded environment, and the folding process is aided by a range of auxiliary proteins (3, 4). These proteins include molecular chaperones, whose main role is to protect the incompletely folded polypeptide chain from non-productive interactions, particularly those that result in aggregation, and folding catalysts, whose job is to speed up potentially slow steps in the folding process such as those associated with the isomerization of peptidylprolyl bonds and the formation of disulphide linkages. It is evident, however, that the code for folding is contained within the amino-acid sequence of the protein itself because it has been shown that proteins can reach their correct folded structure *in vitro* in the absence of any auxiliary factors, providing that appropriate conditions can be found (5). The questions of how the fold is encoded in the sequence, and how the process of folding takes place, are at last beginning to be answered in a credible manner. Progress in this area has come about as a consequence of novel experimental strategies to probe the structural transitions that take place during folding *in vitro,* and of innovative theoretical studies designed to simulate these transitions (6, 7, 8). Perhaps of greatest importance has been the fact that these approaches have been brought together in a synergistic manner to advance our fundamental understanding of this highly complex process (7, 8, 9).

Our present understanding of the folding *in vitro* of small proteins, typically those of less than about 100 residues, is that the rate of folding is limited primarily by the time required to find the crucial interactions that are needed to permit rapid progression to the native structure. For larger proteins, however, the folding process is typically slower and more complex, and is usually associated with the population of one or more partially folded intermediate states. In addition, events that may be termed misfolding may take place during the search for the stable native-like contacts between residues (7). That such complexities are seen even in the benign environment of a dilute solution of a pure protein suggests that they are even more likely to occur in the crowded environment of the cell. Undoubtedly, molecular chaperones are able to mitigate some of the consequences of this complex behaviour and provide some protection for the incompletely folded chain (4). But the idea that proteins can misfold, or fold to intermediates that may undergo undesirable side reactions such as aggregation, provides insight into potential problems that can arise during folding even in the best designed environments. Folding and unfolding are also now known to be coupled to many of the key events in the functioning of a biological system, including translocation of proteins across membranes, protein trafficking, secretion of extracellular proteins, and the control and regulation of the cell cycle (10). Thus, the failure of proteins to fold, or to remain folded under physiological conditions, is likely to cause malfunctions and hence disease. Indeed, an increasing number of diseases is now linked to phenomena that can loosely be described as misfolding; a selection of these is given in table 1.

Table 1. Representative protein folding diseases. (ER, endoplasmic reticulum. Data from (11), (14), (15), (16)).

Disease	Protein	Site of folding
hypercholesterolaemia	low-density lipoprotein receptor	ER
cystic fibrosis	cystic fibrosis transmembrane regulator	ER
phenylketonuria	phenylalanine hydroxylase	cytosol
Huntington's disease	huntingtin	cytosol
Marfan syndrome	fibrillin	ER
osteogenesis imperfecta	procollagen	ER
sickle cell amaemia	haemoglobin	cytosol
α1-antitrypsin deficiency	α-1-antitrypsin	ER
Tay-Sachs disease	β-hexosaminidase	ER
scurvy	collagen	ER
Alzheimer's disease	β-amyloid/presenilin	ER
Parkinson's disease	α-synuclein	cytosol
scrapie/Creutzfeldt- Jakob disease	prion protein	ER
familial amyloidoses	transthyretin/lysozyme	ER
retinitis pigmentosa	rhodopsin	ER
cataracts	crystallins	cytosol
cancer	P53	cytosol

PROTEIN AGGREGATION AND AMYLOID DISEASES

Among the diseases listed in table 1 are those that are associated with the deposition of proteinaceous aggregates in a variety of organs such as the liver, heart and brain (14, 15, 16, 17). Many of these diseases are described as "amyloidoses", because the aggregated material stains with dyes such as Congo red in a manner similar to starch (amylose), and the typical fibrous structures as "amyloid fibrils". A list of known amyloid diseases is given in table 2, along with the protein component that is associated with the extracellular aggregates formed in each case (18).

It is evident that these diseases include many of the most debilitating conditions in modern society, particularly those associated with ageing such as type II diabetes and Alzheimer's disease. Some are familial, some are associated with medical treatment (e.g. haemodialysis) or infection (the prion diseases), and some are sporadic (e.g. most forms of Alzheimer's). Many of the diseases (such as the amyloidoses associated with the protein transthyretin) can be found in both sporadic and familial forms. In addition to these diseases there are others, notably Parkinson's and Huntington's diseases (16, 19), that appear to involve very similar aggregates but which are intracellular not extracellular and are not therefore included in the strict definition of amyloidoses.

Table 2. Fibril protein components and precursors in amyloid diseases. (Data from (21)).

clinical syndrome	fibril component
Alzheimer's disease	Aβ peptide, 1-42, 1-43
spongiform encephalopathies	full length prion or fragments
primary systemic amyloidosis	intact light chain or fragments
secondary systemic amyloidosis	76-residue fragment of amyloid A protein
familial amyloidotic polyneuropathy I	transthyretin variants and fragments
senile systemic amyloidosis	wild-type transthyretin and fragments
hereditary cerebral amyloid angiopathy	fragment of cystatin-C
haemodialysis-related amyloidosis	β2-microglobulin
familial amyloidotic polyneuropahty II	fragments of apolipoprotein AI
Finnish hereditary amyloidosis	71-residue fragment of gelsolin
type II diabetes	fragment of islet-associafed polypeptide
medullary carcinoma of the thyroid	fragments of calcitonin
atrial amyloidosis	atrial natriuretic factor
lysozyme amyloidosis	full length lysozyme variants
insulin-related amyloid	full length insulin
fibrinogen α-chain amyloidosis	fibrinogen α-chain variants

Among the proteins linked with amyloidosis is lysozyme, a protein whose folding we have studied in particular depth (7, 10). Our studies led to the idea that this protein would be an exciting one to choose to try to understand at the molecular level the nature of the "misfolding" transition that converts the protein from a soluble to a fibrillar structure (20). One of the striking characteristics of the amyloid diseases is that the fibrils associated with all of them are very similar in their overall properties and appearance (18). The fibrils are typically long (often several microns), unbranched and ca. 10 nm in diameter. They have a variety of tinctorial properties, notably staining with Congo red and exhibiting a green birefringence under polarized light. A range of experiments, particularly X-ray fibre diffraction, indicates that the fibrils have extensive b-sheet character, and that these sheets run perpendicular to the fibril axis to generate what is described as a cross-b structure (18). This observation is remarkable in view of the fact that the soluble native forms of the proteins associated with these diseases vary considerably in their size and secondary structure. Moreover, some of the proteins are intact in the fibrous form whilst others are at least partially degraded. This similarity of the fibrillar forms of the proteins prompted the proposal that there are strong similarities in the inherent structure of the amyloid fibrils and in the mechanism by which they are formed (12, 18). Thus the study in depth of the relationship between the folding and "misfolding" of one system could have very general value in understanding this whole class of diseases.

One of the very important observations in this regard is that the fibrillar forms of many of the disease-related proteins can be generated *in vitro* from the normal soluble forms. In the case of fibrils formed from peptides (including fragments of larger proteins) that are largely unstructured in solution, such fibrils typically form under a wide range of solution conditions. In the case of fibrils formed from intact globular proteins, however, the fibrils typically form under conditions under which the native state is significantly destabilized (15, 27). Thus in the case of the two known disease-related human lysozyme variants, fibrils form most readily at low pH or at elevated temperatures (20, 23) (see Figure 1).

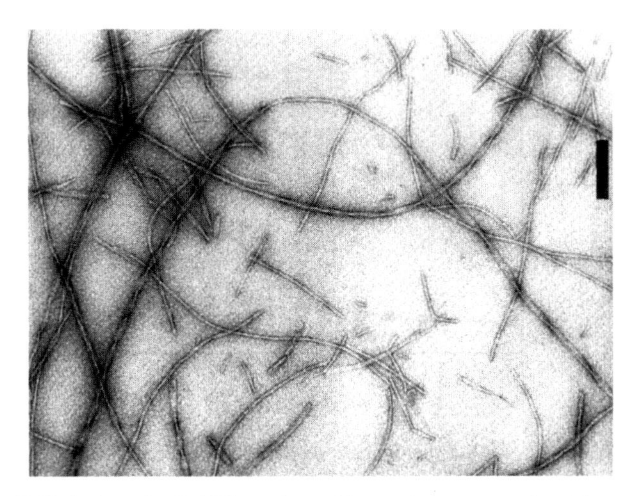

Figure 1. Amyloid fibrils from the Ile56Thr variant of human lysozyme produced by transmission electron microscopy. Scale bar, 200nm. (From (23)).

Experiments to examine the nature of the amyloidogenic variants (Ile56Thr and Asp67His) show that the structures of the proteins in their soluble native states are similar to that of the wild-type protein and have no obvious perturbations that could explain their tendency to aggregate (20). But experiments reveal that the two variants are destabilized relative to the wild-type protein to similar extent although the origin of this instability is different (24). Thus the Ile56Thr variant is destabilized largely because its folding rate is reduced, whilst the Asp67His variant is destabilized largely because it unfolds more rapidly. It therefore appears that the decreased protein stability rather than the altered folding kinetics is a common feature of these two variants. In further experiments it has been demonstrated that the lower stability of the native state results in the population of a partially folded state that is very similar to the major (a-domain) intermediate populated on the folding pathway of the wild-type protein (24, 25). This finding can be rationalized because the mutations destabilizing the native fold are located in the b-domain of the protein, the region that is not highly structured in the predominant intermediate.

This observation suggests a mechanism for the formation of amyloid fibrils from the variant lysozymes, in which the partially folded intermediates aggregate as the first step in the formation of the ordered structures found in the fibrils (20, 23, 24). Calculations based on hydrogen exchange protection measurements suggest that the population of partially folded proteins under physiological conditions is about 1000 times greater in the variants than in wild-type lysozyme (24, 25). This conclusion allows one to speculate that the amyloidogenic variants have sufficient stability to fold efficiently so as to escape the quality control mechanisms in the endoplasmic reticulum and to be secreted into extracellular space (12). However, unlike the wild-type protein, they have insufficient stability to remain in their native states under all conditions to which they are exposed. Moreover, it has been speculated that endosomal compartments where the pH is reduced might be important in the formation of amyloid deposits. Under low pH conditions *in vitro* conversion to amyloid fibrils has been found to be particularly facile (23). In addition, *in vitro* experiments have shown that fibril formation is accelerated substantially when solutions are seeded with preformed fibrils. Such a mechanism has been suggested as being responsible for the rapid onset of some amyloidoses, and indeed of the infectivity of the prion diseases (16).

The Generic Nature of the Amyloid Structure

In studies of the conformation of the SH3 domain from bovine PI3 kinase at low pH, when the protein is in a largely unfolded state, it was found that the protein readily formed a viscous gel (26). Examination of the gel using electron microscopy revealed the presence of large numbers of fibrils that closely resemble those formed from the proteins associated with amyloid diseases. Moreover, the aggregates showed all the other characteristics of amyloid fibrils, and were to all intents and purposes identical to these other structures. This observation prompted us to explore the possibility that similar fibrils could be formed from other proteins by placing them under mildly denaturing conditions that do not immediately result in visible precipitation, and examining the solutions over often prolonged periods of time (27). For a range of representative proteins with no known connection with any disease we have been able to find conditions under which conversion occurs into fibrils very similar to those associated with amyloid disease. We shall refer to these types of structures as "amyloid fibrils" in future, regardless of whether or not they are associated with disease. The proteins studied included wild-type human lysozyme, which forms fibrils under similar but more destabilizing conditions than the amyloidogenic intermediates (23), and the archetypal globular protein, myoglobin, which readily forms fibrils when the heme group is removed (13). For myoglobin it is particularly evident that the protein has undergone a substantial conversion from its soluble a-helical form to the aggregated b-sheet conformation found in the fibrils. Such findings prompted us to conclude that the ability to form amyloid fibrils is not a characteristic associated wholly or primarily with those proteins found to be associated with amyloidoses, but a property that could be common to many or indeed all proteins under appropriate conditions (12, 26, 27, 28).

Models of the structure of amyloid fibrils indicate that the core regions of the protofilaments are based on hydrogen bonds between the polypeptide main chain (18, 29). As the main chain is common to all polypeptides, it explains how the fibrils from different proteins appear so similar, regardless of the length and sequence of the polypeptide involved. In contrast to the situation in native proteins, we suggest that the side chains are not a strong influence on the basic structure of the protofilaments (30). Nevertheless, the manner in which the protofilaments pack together to form mature fibrils may well depend significantly on those parts of the polypeptide chain that are not involved directly in the close-packed β-strands (31). Thus, the fibrils from different peptides and proteins are variations on a common theme. The ability of natural proteins to form amyloid structures does not violate the crucial hypothesis that a protein sequence codes for a single fold (5). The nature of the amyloid core structure is that it is not coded for by the sequence, as it is formed as a consequence of interactions involving the common polypeptide backbone of all proteins. Its rate and ease of formation will of course depend on the sequence, both as a consequence of the readiness for different side-chains to pack together within the structure, and as a consequence of the solubility and stability of the sequence in solution (32). It is the side-chains, however, that code for the specific fold of globular proteins by their ability to pack together in a unique manner to form compact globular structures.

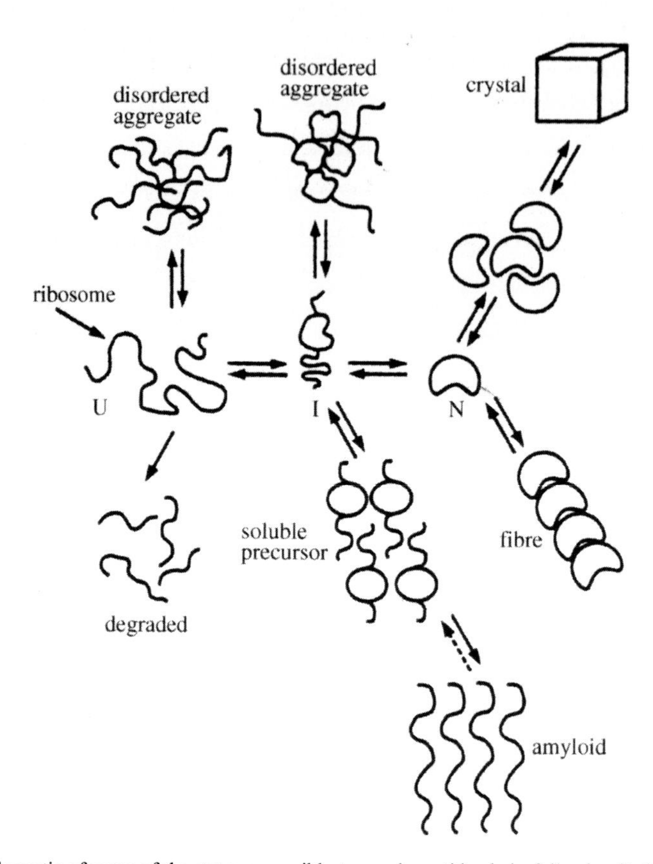

Figure 2. Schematic of some of the states accessible to a polypeptide chain following its biosynthesis. In its monomeric state, the protein is assumed to fold from it highly disordered unfolded state (U) through a partially structured intermediate (I) to a globular native state (N). The native state can form aggregated species, the most ordered of which is a three-dimensional crystal, whilst preserving its overall structure. The unfolded and partially folded states can form aggregated species that are frequently disordered, but highly ordered amyloid fibrils can form through a nucleation and growth mechanism. (From (33)).

The proposal that amyloid fibrils represent a generic structure of polypeptide chains has stimulated us to suggest that the conformational properties of all proteins should be considered in terms of the multiple states that are accessible to such structures (6, 33). This suggestion is illustrated in Figure 2 in a schematic manner. This diagram suggests that the various fates awaiting a polypeptide chain once it has been synthesized in the cell will depend on the kinetics and thermodynamics of the various equilibria between different possible states. Thus, the normal folding process may pass through partially folded states on the route to the fully native state, but the aggregation of these species will be minimized by the presence of molecular chaperones.

In addition, if the protein is able to fold rapidly, any partially folded species will have a short lifetime, reducing the probability of intermolecular interactions occurring. Moreover, once folded, the native state is generally a highly compact structure that conceals the polypeptide main chain within its interior. Such a state is protected from aggregation except through the relatively weak interactions of surface side chains and is unable to form the strong intermolecular hydrogen bonds associated with the polypeptide backbone. Provided that the native state is maintained under conditions where it remains folded, aggregation to amyloid fibrils will be resisted by the kinetic barrier associated with unfolding, even if the aggregated state is thermodynamically more stable. Importantly, the cooperative nature of protein structures means that virtually none of the polypeptide chain in individual molecules is locally unfolded, and that virtually no molecules in an ensemble are globally unfolded, even though native proteins are only marginally stable relative to denatured ones under normal physiological conditions (6, 33).

A Common Origin of Amyloid Diseases

This picture of the various structures accessible to polypeptide chains enables us to speculate on the origins of the amyloid diseases from the point of view of the physicochemical properties of protein molecules. If the stability or cooperativity of the native state of a protein is reduced, for example by a mutation, the population of non-native states will increase, as discussed above for the amyloidogenic variants of lysozyme. This rise will increase the probability of aggregation, as the concentration of polypeptide chains with at least partial exposure to the external environment will be greater. Whether or not aggregation does occur will depend on the concentration of protein molecules, the intrinsic propensity for a given sequence to aggregate when unfolded, and on the rate of the aggregation process. The fact that formation of ordered amyloid fibrils can be seeded, like the well-studied processes of crystallization and gelation, means that once the aggregation process is initiated it often proceeds very rapidly (23, 34, 35). In the absence of seeding there can be long "lag" phases before aggregation occurs. This lag can be thought of as arising because the growth of a fibril cannot occur until a "nucleus" of a small number of aggregated molecules is formed. Such a nucleus can be formed by the local fluctuations in concentration that occur in solution as a result of random molecular motion. When such fluctuations result in a local concentration of molecules above a critical value, the molecules associate with one other to form a species that is sufficiently large to have intrinsic stability, and hence to grow in size by interacting with other molecules in the solution. The act of seeding provides such nuclei to the solution and hence reduces or abolishes the lag phase.

On this view of the aggregation process, the critical first step for globular proteins is the partial or complete unfolding of the native structure. In the case of most proteins, except perhaps the smallest ones, unfolding under physiological conditions will not generate the type of highly unfolded states seen in high concentrations of denaturant. Instead the denatured protein will be more stable in a partially collapsed state that may well resemble intermediates observed in the normal folding process (36). The generic nature of the structure and mechanism of amyloid formation suggests that the nature of the residual structure in such intermediates has little direct importance in dictating the structure of the resulting aggregates. It may, however, it may indicate which regions of the protein are most likely to be incorporated in the b-sheet segments of the fibrils. It is, however, the enhanced ability of some mutated proteins to access partially or completely unfolded states that is the underlying origin of many of the familial amyloid diseases such as those involving lysozyme that are discussed above (20, 25). Moreover, in some cases mutational changes can also enhance the propensity of unfolded or partially folded states to aggregate, providing an additional mechanism for the amyloidogenity of some disease-related variants (37, 38).

This general view of amyloid formation can readily be extended to include the existence of sporadic as well as familial and infectious diseases. Sporadic diseases could arise from the loss of the normal control and regulation processes that enable proteins to be maintained in their required states under all conditions in living organisms organism (12, 27). It is perhaps particularly likely that such control is lost in ageing, and the majority of the cases of sporadic diseases such as Alzheimer's or type II diabetes are associated with old age. It is significant that in a high proportion of elderly people even wild-type transthyretin, which in its mutant forms is associated with familial amyloidosis, is found as amyloid structures in organs such as the heart (15). The exact manner in which this happens is unclear, but it could be as the result of statistical factors (comparable with those observed in lag phases *in vitro*) or due to the effects of changes in the cellular environment, or the failure of the normal degradation mechanisms for proteins. Interestingly, many of the diseases in this category involve the deposition of peptide fragments, and the process of degradation in compartments such as lysozomes involves conditions such as low pH that serve to unfold proteins prior to the action of proteases. Such mildly denaturing conditions are particularly favourable for the nucleation and growth of amyloid structures.

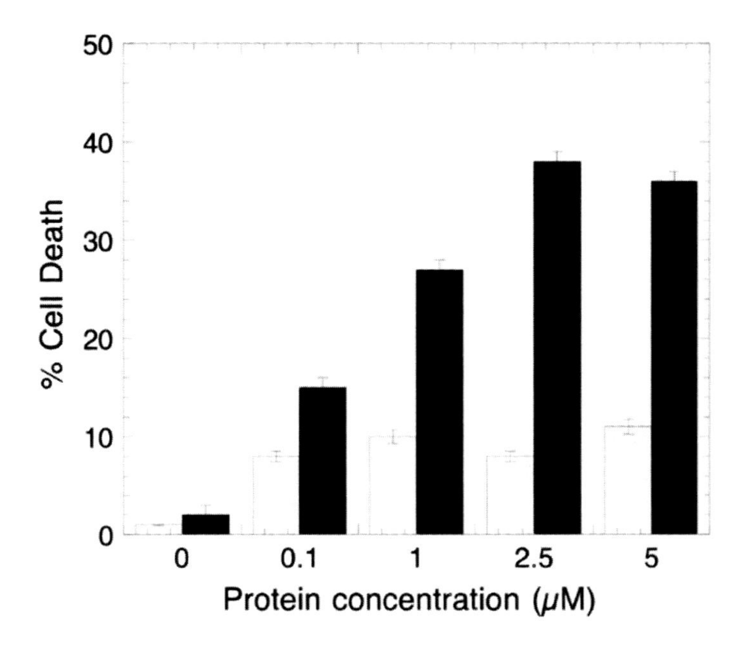

Figure 3. Percentage of cell deaths induced by 48-h-aged Hyp-F-N aggregates at different protein concentrations. Solid bars refert to aggregated protein, dotted bars to control experiments performed in the presence of soluble protein.

CONCLUDING REMARKS

It seems likely that biological evolution has resulted in the selection of polypeptide sequences that are able to fold to compact, globular and soluble forms that resist aggregation and conversion to fibrillar structures, at least when protected in a highly stable and controlled environment (12). These remarkable structures are the native states of proteins that are involved in every process occurring in the cell. The appearance of amyloid deposits in living systems may therefore be associated with mutations that destabilize proteins sufficiently for them to convert into fibrils when the wild-type protein would not, but leave them sufficiently stable to evade the quality control mechanisms in the cell and to function sufficiently normally to allow the organism to develop and reproduce. Amyloid deposits also appear in old age where evolutionary pressure is reduced after the reproductive life span, and in other conditions such as kuru or bovine spongiform encephalopathy (BSE), which are connected with abnormal practices such as ingestion of tissue from other members of the same species (17).

The generic picture of amyloid structure and the mechanism of its formation therefore provides a conceptual framework for linking together the various pathological conditions associated with deposition of this material. This hypothesis has recently been extended by the observation that the species formed *in vitro* during the early stages of aggregation of proteins not linked to disease, like the early aggregates of disease associated proteins, can be highly toxic when added to cells in culture (39). This finding suggests that the control and regulation of aggregation can be even more crucial to the viability of the organism than was previously thought if misfolded proteins are not just non-functional but also potentially toxic. Moreover, it suggests that there may be at least common features in the pathology of the various misfolding diseases. Our knowledge of the mechanism of fibril growth and degradation should therefore help enormously to design effective strategies for the prevention or treatment of the various members of the family of amyloid diseases.

Moreover, the fundamental knowledge that is emerging from our ability to probe the aggregation properties of a range of sequences, rather than just those identified in association with recognized diseases, should enable the role of particular interactions and specific structural motifs in these processes to be explored (22, 27, 38). This knowledge could result in future approaches to drug design that are significantly more general and potentially more effective than those currently being explored. Given that many of the amyloid diseases are associated with old age, and modern medical and agricultural practices, the need for novel approaches to the avoidance or treatment of these conditions will be increasingly important in the future. One can speculate that as the human life span increases, and the complexity of our societies increases, the number of cases of the known diseases is likely to increase substantially, and indeed that novel diseases associated with the aggregation of proteins not so far linked to clinical symptoms might become highly significant in future years.

This article is an edited and updated version of a paper first published in Phil. Trans. R. Soc. Lond. B, 356, 133-145 (2001). The research of C.M.D is supported in part by a programme grant from the Wellcome Trust.

REFERENCES

[1] Dobson, C. M. & Fersht, A. R. (eds.) (1995). Protein folding. *Phil. Trans. R. Soc. Lond. B* **348**:1-119.

[2] Pain, R. H. (ed.) (2000). Mechanisms of Protein Folding, Second Edition Oxford.

[3] Gething, M.-J. & Sambrook, J. (1992). Protein folding in the cell. *Nature* **355**:33-45.

[4] Ellis, R. J. & Hartl, F. U. (1999). Principles of protein folding in the cellular environment. *Curr. Opin. Struct. Biol.* **9**: 102-110.

[5] Anfinsen, C. B. (1973). Principles that govern the folding of protein chains. *Science* **181**:223-230.

[6] Dobson, C. M. & Karplus, M. (1999). The fundamentals of protein folding: bringing together theory and experiment. *Curr. Opin. Struct. Biol.* **9**:92-101.

[7] Dinner, A. R., Sali, A., Smith, L. J., Dobson, C. M., Karplus, M. (2000). Understanding protein folding via free energy surfaces from theory and experiment. *Trends Biochem. Sci.* **25**:331-339.

[8] Daggett. V. & Fersht. A. R. (2002). *Cell.*

[9] Vendruscolo, M., Paci, E., Dobson, C. M., Karplus, M. (2001). Three Key Residues Form a Critical Contact Network in a Transition State for Protein Folding, *Nature* **409**: 641-646.

[10] Radford, S. E. & Dobson, C. M. (1999). From computer simulations to human disease: emerging themes in protein folding. *Cell* **97**:291-298.

[11] Thomas, P. J., Qu, B. H., Pederson, P. L. (1995). Defective protein folding as a basis of human disease. *Trends Biochem. Sci.* **20**:456-459.

[12] Dobson, C. M. (1999). Protein misfolding, evolution and disease. *Trends Biochem. Sci.* **24**:329-332.

[13] Fandrich, M., Fletcher, M. A., Dobson, C. M. (2001). Formation of amyloid fibrils from myoglobin. (Submitted).

[14] Tan, S. Y. & Pepys, M. B. (1994). Amyloidosis. *Histopathology* **25**:403-414.

[15] Kelly, J. W. (1998). The alternative conformations of amyloid proteins and their multi-step assembly pathways. *Curr. Opin. Struct. Biol.* **8**:101-106.

[16] Lansbury, P. T. (1999). Evolution of amyloid: what normal protein folding may tell us about fibrillogenesis and disease. *Proc. Natl. Acad. Sci. USA* **96**:3342-3344.

[17] Prusiner, S. (1997). Prion diseases and the BSE crisis. *Science* **278**:245-251.

[18] Sunde, M. & Blake, C. C. F. (1997). The structure of amyloid fibrils by electron microscopy and X-ray diraction. *Adv. Protein Chem.* **50**:123-159.

[19] Perutz, M. F. (1999). Glutamine repeats and neuro-degenerative disease: molecular aspects. *Trends Biochem. Sci.* **24**:58-63.

[20] Booth, D. R. (and 10 others) (1997). Instability, unfolding and aggregation of human lysozyme variants underlying amyloid fibrillogenesis. *Nature* **385**:787-793.

[21] Sunde, M., Serpell, L. C., Bartlam, M., Fraser, P. E., Pepys, M. B., Blake, C. C. F. (1997). Common core structure of amyloid fibrils by synchrotron X-ray diffractions. *J. Mol. Biol.* **273**:729-739.

[22] Chiti, F., Taddei, N., White, P. M., Bucciantini, M., Magherini, F., Stefani, M., Dobson, C. M. (1999). Mutational analysis of acylphosphatase suggests the importance of topology and contact order in protein folding. *Nature Struct. Biol.* **6**:1005-1009.

[23] Morozova-Roche, L. A., Zurdo, J., Spencer, A., Noppe, W., Receveur, V., Archer, D. B., Joniau, M., Dobson, C. M. (2000). Amyloid fibril formation and seeding by wild type human lysozyme and its disease related mutational variants. *J. Struct. Biol.* **130**:339-351.

[24] Canet, D., Sunde, M., Last, A. M., Miranker, A., Spencer, A., Robinson, C. V., Dobson, C. M. (1999). Mechanistic studies of the folding of human lysozyme and the origin of amyloido-genic behavior in its disease related variants. *Biochemistry* **38**:6419-6427.

[25] Canet, D., Last, A. M., Tito, P., Sunde, M., Spencer, A., Archer, D. B., Redfield, C., Robinson, C. V., Dobson, C. M. (2002). Local Cooperativity in the Unfolding of the Asp67His Variant of Human Lysozyme is a Key Aspect of its Amyloidogenic Behaviour. *Nature Struct. Biol.* **9**:308-315.

[26] Guijarro, J. I., Sunde, M., Jones, J. A., Campbell, I. D., Dobson, C. M. (1998). Amyloid fibril formation by an SH3 domain. *Proc. Natl. Acad. Sci. USA* **95**:4224-4228.

[27] Chiti, F., Webster, P., Taddei, N., Clark, A., Stefani, M., Ramponi, G., Dobson, C. M. (1999). Designing conditions for in vitro formation of amyloid protolaments and fibrils. *Proc. Natl. Acad. Sci. USA* **96**:3590-3594.

[28] Dobson, C. M. (2001). Protein folding and human disease *Phil. Trans. R. Soc. Lond. B.*

[29] Minton, A. P. (2000). Implications of macromolecular crowding for protein assembly. *Curr. Opin. Struct. Biol.* **10**:34-39.

[30] Jimënez, J. L., Guijarro, J. I., Orlova, E., Zurdo, J., Dobson, C. M., Sunde, M., Saibil, H. R. (1999). Cryo-electron micro-scopy structure of an SH3 amyloid fibril and model of the molecular packing. *EMBOJ.* **18**:815-821.

[31] MacPhee, C. E. & Dobson, C. M. (2000). Formation of mixed fibrils reveals the generic nature and potential application of protein amyloid structures. *J. Am. Chem. Soc.* **122**:12707-12713.

[32] Chamberlain, A. K., MacPhee, C. E., Zurdo, J., Morozova-Roche, L. A., Hill, H. A. O., Dobson, C. M., Davis, J. J. (2000). The ultrastructural organisation of amyloid brils by atomic force microscopy. *Biophys. J.* **79**:3282-3293.

[33] Chiti, F., Capanni, C., Taddei, N., Stefani, M., Ramponi, G., Dobson, C. M. (2000). Specific regions of a protein determine the kinetics of aggregation and amyloid formation (in preparation).

[34] Dobson, C. M. (1999). How do we explore the energy landscape for folding? In: Simplicity and complexity in proteins and nucleic acids (ed. H. Fraunfelder, J. Deisenhofer & P. G. Wolynes), pp. 15-37. Berlin: Dahlem University Press.

[35] Harper, J. D. & Lansbury, P. T. (1997). Models of amyloid seeding in Alzheimer's disease and scrapie: mechanistic truths and physiological consequences of the time-dependent solubility of amyloid proteins. *A. Rev. Biochem.* **66**, 385-407.

[36] Krebs, M. R. H., Wilkins, D. K., Chung, E.W., Pitkeathly, M. C., Chamberlain, A., Zurdo, J., Robinson, C. V., Dobson, C. M. (2000). Formation and seeding of amyloid fibrils from wild-type hen lysozyme and a peptide fragment from the b-domain. *J. Mol. Biol.* **300**:541-549.

[37] Dobson, C. M., Sali, A., Karplus, M. (1998). Protein folding: a perspective from theory and experiment. *Angew. Chem. Int. Ed. Eng.* **37**:868-893.

[38] Villegas, V., Zurdo, J., Filimonov, V. V., Aviles, F. X., Dobson, C. M., Serrano, L. (1999). Protein Engineering as a Strategy to Avoid Formation of Amyloid Fibrils, *Protein Sci.* **9**:1700-1708.

[39] Chiti, F. Taddei, N. Baroni, F. Capanni, C. Stefani, M. Ramponi G., Dobson, C. M. (2002). Kinetic Partitioning of Protein folding and Aggregation, *Nature Struct. Biol.* **9**:137-143.

[40] Bucciantini, M., E. Giannoni, F. Chiti, F. Baroni, L. Formigli, J. Zurdo, N. Taddei, G. Ramponi, C. M. Dobson, Stefani, M. (2002). Inherent Cytotoxicity of Aggregates Implies a Common Origin for Protein Misfolding Diseases. *Nature* **416**:507-511.

Using Evolutionary Information to Study Protein Structure

Richard A. Goldstein

Siena Biotech, via Fiorentina 1, 53100 Siena, Italy
E-Mail: rgoldstein@sienabiotech.it

Received: 26th July 2002 / Published: 15th May 2003

Abstract

The genomic data available to computational biologists represents the product of the complex processes of evolution. In particular, the forces of mutation, duplication, and selection have acted to sculpt modern protein sequence and structure in the context of changing functional requirements. Just as crystallographers are able to determine protein structures through an analysis of X-ray diffraction patterns, we wish to read the evolutionary history of proteins in order to understand their structures, functions, and interactions. To this end, we have been developing models of natural site substitutions that are informed by the protein structure and function and the resulting variations in selective pressures, even when the structure and function of the protein are unknown. By phrasing the substitution process in terms of the underlying properties of the constituent amino acids we can build models that are both much more accurate and more interpretable. The model is applied to a large set of globular proteins as well as a set of G-protein coupled receptors, identifying general structural and functional features of these biomolecules.

Introduction

The various genome projects have produced a plethora of gene sequences encoding proteins for which we have little information. While there are extensive experimental efforts to characterize the structure, function, and other characteristics of these proteins, there still remains a substantial backlog. In addition, many proteins of major interest are resistant to many of these experimental techniques. This has helped to spur the development of techniques to predict the characteristics of these proteins based only on sequence information. Often we have multiple sequences of related proteins from different organisms.

It has been long recognized that these multiple sequences provide us with a valuable opportunity, that a set of related sequences convey more information than just a single example. The challenge has been to extract meaningful information from these multiple sets.

Given an alignment, it is possible to identify the conserved residues, to characterize the amount of sequence variation at each location using such concepts as "sequence entropy", and to look for correlated changes between different locations in the protein. These approaches generally treat the observed protein sequences as a random sampling from the space of all possible sequences. This, of course, is false. One obvious problem is the uneven distribution of proteins among different organisms, depending upon the relative importance of the species to individual scientific investigators. More insidiously, homologous proteins are related by a phylogenetic structure that can induce confounding tendencies in the data. For example, Figure 1 shows a phylogenetic pattern where two substitutions occurred in different branches of the tree. In this simple example, there is a complete correlation between the third and seventh positions, even though this does not represent the effect of compensatory substitutions. One approach to handling these complications is to model the evolutionary process explicitly. This is the approach that we take here.

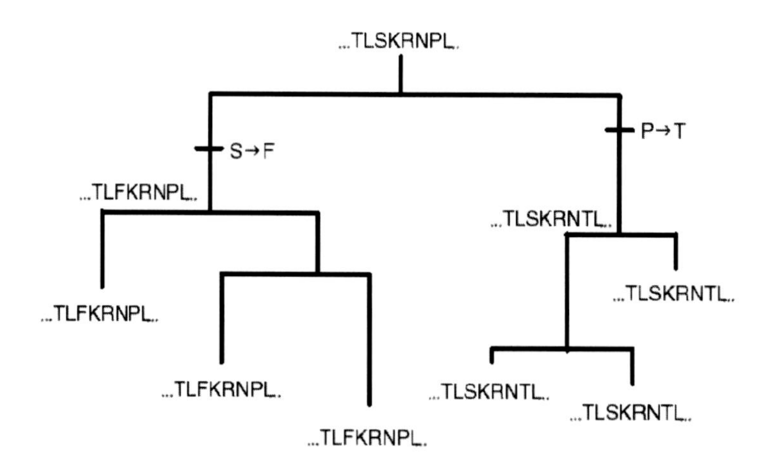

Figure 1. Example of an evolutionary trajectory producing an artificial correlation between sequence locations.

The standard method to model the site substitutions that occur during evolution is through a "substitution matrix", a 20 x 20 matrix representing the probability that one amino acid would be replaced by another in a given length of evolutionary time. Standard approaches generally use a single substitution matrix for all locations in the protein, implicitly assuming that all locations in the protein can be represented by the same model, that is, are under similar selective pressure. This is, of course, unrealistic. It has been shown that substitution rates vary with surface accessibility, secondary structure, and functional significance. One method to approach this problem is to subdivide the various locations in proteins according to their local structure, constructing and using structure-dependent substitution matrices (1, 2, 3, 4). This approach still assumes that all locations with the same local structure are under similar selective pressure, ignoring differences based on the inevitable "coarse graining" of the structural classifications as well as selective pressure due to function.

Recently we developed an approach, which we call a Hidden States Model (HSM), for dealing with these differences in selective pressure (5, 6, 7, 8). In this model, each location is assumed to belong to one of a set of possible "site classes", each corresponding to a single substitution matrix. The various substitution matrices are unknown, as is the site class to which each location belongs. Instead, each location in the protein has the same set of a priori probabilities for belonging to each site class. The a priori probabilities as well as the set of substitution matrices are determined based on a set of related proteins through a maximum likelihood formulation. The result of this procedure is a set of site classes with corresponding substitution matrices, as well as the ability to calculate the a posteriori probability that any given location is a member of any particular site class. This then provides us with information regarding 1) which locations are under related selective pressure, 2) what is the nature of this selective pressure, and 3) when is the selective pressure different for different subsets of proteins.

The central challenge in this approach is the total number of parameters that must be adjusted in the optimization process. We deal with this situation by representing the entire substitution matrix with a biologically-inspired reduced set of parameters. In general, we consider the local propensity, or fitness, of each amino acid for any location described by a given site class, and then represent the probability of substitution of one amino acid for another in terms of the differences in these local fitnesses. The functional forms of this representation can be quite general, with additional parameters that can be optimized based on the observed data.

In this paper, we first investigate the nature of the substitution process at different types of locations in a set of globular proteins. We then demonstrate the application of these models for understanding the selective pressure acting on one particular set of proteins, G-protein coupled receptors (GPCRs).

METHODS

We first recap our model for site substitutions, as described elsewhere (5, 7, 8). We first consider that there are a number of different site classes, which characterize locations in the protein under similar selective pressure. As described above, the model does not assign locations to site classes; instead we define an unknown prior probability $P(k)$, that any given location belongs to site class k. As all locations must belong to a site class, $\Sigma_k P(k) = 1$.

We need to reduce the number of adjustable parameters that characterize each particular substitution matrix. In order to do this, we first consider that there is a relative fitness $F_k(A_i)$ of amino acid A_i for any location described by a particular site class k, related to the logarithm of the probability of finding such an amino acid at this location described by this site class. The instantaneous rate of substitution Q^k_{ij} from amino acid A_i to A_j at site class k is then reflected by the relative changes in fitness. In this paper, we use a few different models. In our analysis of a general set of globular proteins, we use so-called Metropolis kinetics, where advantageous substitutions ($\Delta F \equiv F_k(A_j) - F_k(A_i) \geq 0$) are accepted at a maximum site-class dependent rate ν_k, while disadvantageous substitutions ($\Delta F < 0$) are accepted with a probability that decreases exponentially with the resulting change in fitness.

$$Q^k_{ij} = \left\{ \begin{array}{c|c} \nu_k e^{\Delta F} & \Delta F < 0 \\ \nu_k & \Delta F \geq 0 \end{array} \right\} \quad (1)$$

The Metropolis scheme is the only kinetics scheme ensuring detailed balance, and where a favorable substitution is always accepted at the maximum rate.

In addition, as we were most interested in modeling the general nature of the selective pressure at different locations, we further parameterized the fitness of each amino acid at a given site class as a function of the physical-chemical properties of the amino acids:

$$F_k(A_i) = \sum_l \alpha^l_k (q_l(A_i) - \phi^l_k)^2 \quad (2)$$

where $q_l(A_i)$ represents the value of physical-chemical property l of amino acid A_i, and α^l_k and ϕ^l_k represent site-class specific adjustable parameters. In this study, we used the four orthogonal property indices developed by Scheraga and coworkers (9). The first property is positively correlated with turn propensity and negatively correlated with α-helix propensity; the second is positively correlated with size and bulk, the third is positively correlated with β-sheet propensity, and the last is negatively correlated with hydrophobicity, meaning hydrophilic residues have high positive values in this index.

For the analysis of the G-Protein Coupled Receptors, we used a more general function of the form:

$$Q_{ij}^{k} = v_k \, e^{-\lambda_k (\Delta F)^2} \frac{\beta_k e^{\Delta F/2}}{\beta_k e^{|\Delta F/2|} + e^{-|\Delta F/2|}} \quad (3)$$

where ν_k again characterizes the overall substitution rate for site class k, and λ_k and β_k are parameters of the function. Note that this model is equivalent to the Metropolis scheme under the conditions $\lambda_k = 0$ and $\beta_k \gg 1$. In contrast to the case for the general set of globular proteins, we left the values of $F_k(A_j)$ as independently adjustable parameters.

To determine the substitution matrix M, representing the possible substitutions from amino acid A_i to A_j for any particular amount of evolutionary time t, the Q matrix is exponentiated:

$$M_{(k)}(t) = e^{tQ_{(k)}} \quad (4)$$

The model involves a large number of adjustable parameters. We will notate the parameters for site class k, including the prior probability $P(k)$, as $\{\theta\}_k$. For the study of the large set of globular proteins, this includes $P(k)$, ν_k, and α^l_k and ϕ^l_k for the four different physical-chemistry parameters (the values of $q_l(A_i)$ for the twenty amino acids are measured, not adjustable, parameters). For the GPCR study, these parameters include $P(k)$, ν_k, λ_k and β_k, and the twenty fitness parameters of the amino acids $F_k(A_i)$ (as the fitness values are relative, one of these parameters is set to zero). We will notate the entire set of all parameters, including the parameters for all of the site classes, as Θ.

These parameters are adjusted in order to maximize the log likelihood, that is, the log of the conditional probability that the observed data would result if the model were correct. At each location l, we first calculate the probability $P(D_l | \theta_k, T)$ of the observed amino acids at that location, D_l, resulting from the evolutionary dynamics if the location was assigned to site class k with model parameters θ_k. given the evolutionary tree topology and branch lengths T. Since each location can be represented by any of the site classes and each site class has distinct parameters θ_k we have to sum over all possible site classes to calculate the total likelihood for that location, L_l:

$$L_l = \sum_k P(D_l \mid \theta_k, T)\, P(k) \quad (5)$$

The log-likelihood for the entire set of proteins is calculated as the sum of the log of this likelihood at each location in the alignment.

While we do not know to which site class a location belongs *a priori*, following optimization of the model we can calculate *a posteriori* probabilities. The conditional probability that a location l belongs to site class k is given by:

$$P(k \mid D_l) = \frac{P(D_l \mid \theta_k)\, P(k)}{\sum_k P(D_l \mid \theta_k)\, P(k)} \quad (6)$$

DATASETS

A general protein data set was constructed by selecting 42 proteins of length greater than 80 residues from the list constructed by Hobohm and Sander (10), all with 6 to 11 homologs of 30% or greater sequence identity listed in the HSSP database (11). The average number of homologs for each protein was 10.5. A multiple alignment and phylogenetic tree was created for each set using the program ClustalV (12). The sequence, structure, and surface accessibilities were found by use of the DSSP program on the corresponding PDB files (13, 14). Residues were considered exposed if greater than 18% of their surface area was exposed to solvent.

Models with two site classes were optimized where $F_k(A_i)$ was a function of all four of Scheraga's orthogonal indices. Separate analyses were performed for buried and exposed residues. In each case, we calculated how much each physical chemical parameter contributed to the variance of the fitness values of the different amino acids for each of the site classes.

For the GPCR project, we selected a group of 185 amine-binding proteins, obtaining the multiple alignment from GPCRdb (15). We used PHYLIP (16), which uses a parsimony approach to calculate the best tree from a given set of data. Resulting trees were optimized for their branch lengths using PAML (17). A model consisting of 5 site classes was optimized. We then calculated the posterior probabilities of the site classes for each location In order to interpret the selective pressure described by each site class, we calculated the correlation coefficient between the fitness values and physicochemical properties of amino acids. These properties were derived from the AAindex database (18), which contains 434 different amino acid indices. We avoided indices related to spectroscopic methods and selected 145 physicochemical indices (see supplement for AAindex database codes of the used indices). Solvent accessibility calculations for rhodopsin were done using the publicly available software GETAREA 1.1 (19).

RESULTS

General properties for globular proteins

Representations for the selective constraints on exposed locations is shown in Figure 2. The optimization resulted in two distinct site classes, one site class representing the majority of sites (represented by the relative size of the pie charts for the two site classes), with a faster rate of variation (larger v_k), with the fitness of the amino acids primarily determined, unsurprisingly, by the hydrophilicity. In addition, there was a preference for small residues as well as a slight preference for residues with high turn propensity. The less common site class, conversely, had a slower rate of variation (smaller v_k), and had a strong preference for hydrophobic residues.

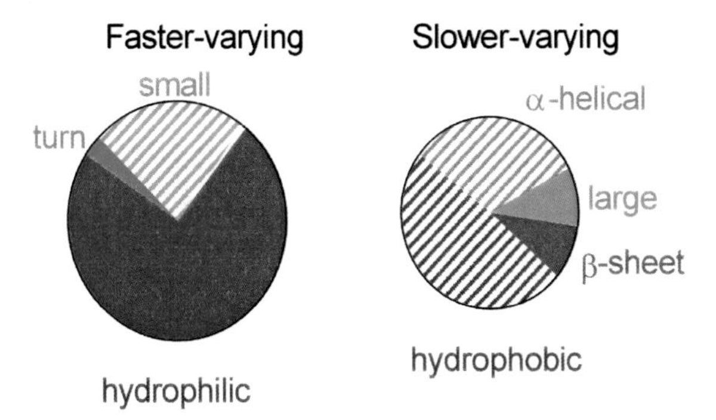

Figure 2. Pie charts representing the various contributions to the selective pressures acting on surface locations belonging to the two site classes. The relative sizes of the charts represents the percentage of the surface locations assigned to these classes. The color scheme represents the various Scheraga factors, including hydrophilicity (red), α-helix or turn propensity (blue), bulk (green), and ß-sheet propensity (magenta). Solid colors represent a positive correlation with the Scheraga factor, while a striped pattern represents a negative correlation.

Helical propensity was also important, with smaller preferences for bulkier residues and residues with a larger sheet propensity.

The situation for buried residues is portrayed in Figure 3. The faster-varying locations actually occupied a minority of the buried locations, and predictably preferred hydrophobic residues, although this preference was less strong than the tendency for faster-varying exposed locations for hydrophilic residues. Equally strong was a tendency for residues with propensity for β sheets, as well as a moderate preference for residues with α-helical preferences, as well as large residues. The larger group of locations were in a slower-varying site class with a strong tendency towards small residues, and smaller preferences for hydrophilicity, turn propensity, and β-sheet propensity.

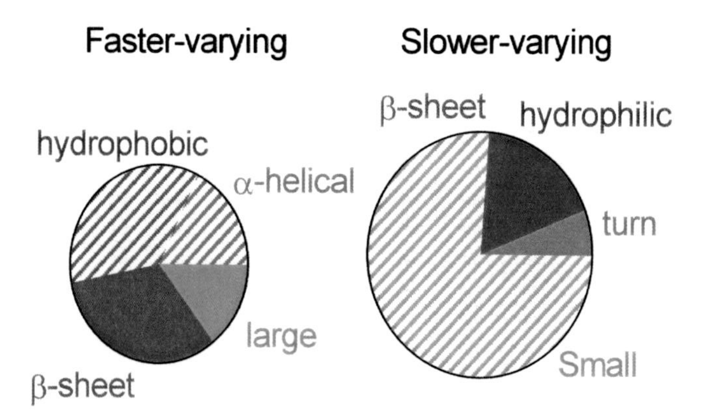

Figure 3. Pie charts representing the various contributions to the selective pressures acting on buried locations belonging to the two site classes. The relative sizes of the charts represents the percentage of the surface locations assigned to these classes. The color scheme is as for Figure 2.

G-Protein Coupled Receptors

The various parameters for the five site-class model obtained for the amine-binding GPCRs are listed in Table 1.

By mapping the locations in the amine-binding proteins to the known structure of Bovine bacteriorhodopsin, we can identify which locations are assigned to different site classes. Locations which are likely to reside in the membrane are largely assigned to site classes 1, 2, and 3, while loop locations are almost entirely assigned to classes 4 and 5. In addition, transmembrane locations in site classes 3 and 4 are generally in areas exposed to the membrane, while locations in site classes 1 and 2 generally face into the interior of the protein.

Table 1. Overall substitution rates and properties of preferred amino acids for the five site-class model optimized on the set of amine-binding GPCRs

site class (k)	Rate	Preferences
1	0.01	alpha-helical propensity
2	0.12	Hydrophobic
3	0.41	Hydrophobic, membrane
4	0.97	Flexible, buried
5	2.64	Polarizable

DISCUSSION

Both the buried and exposed locations can be divided into two different site classes, a faster-varying set of locations and a slower-varying set. It is not surprising that the faster varying locations on the exterior prefer hydrophilic residues, while the faster-varying locations on the interior prefer hydrophobic. It is surprising, however, that for the slower-varying sites in both contexts these preferences are reversed. It is likely that the faster-varying sites are under less purifying selective pressure than the sites that vary more slowly. While most locations in the inside would be under some selective pressure to remain hydrophobic, the other, more specialized forms of selective pressure acting on some locations may favor conservation of such things as particular hydrogen bond or ionic bond formations. In these locations, this specialized selective pressure may well favor hydrophilic residues in a way that "trumps" the more general forms of selective pressure felt by more average locations in the protein. These locations would have slower substitution rates as well as more complex forms of the selective pressure. Similarly, these locations may be involved in specific packing or aromatic stacking interactions, so the preference for larger residues might be explainable. Conversely, the locations on the protein exterior that change slowly might be under more specific forms of selective pressure that prefer hydrophobic residues. In both instances, the needs of stabilizing a specific conformation may result in a tendency for specific locations to have selective pressure opposite in form to that of other, seemingly similar locations. One interesting point to note is that, for buried locations, the majority of sites are slower varying. Another observation is that the dominant hydrophobic selective pressure is on faster-varying external locations to remain hydrophilic. This provides further evidence for the reverse hydrophobic effect, that is, the need to avoid stabilizing alternative conformations where these particular locations are buried (20).

The analysis of the GPCRs demonstrate that we can obtain specific structural information from sets of aligned sequences, even identifying trans-membrane residues facing into the protein interior or out into the lipid membrane. All this information is gathered as a result of the optimization procedure, with no *a priori* knowledge about structure or function. As such, it is a powerful way of generating important information about the new proteins whose sequences are becoming available.

ACKNOWLEDGMENTS

This contribution represents the work of a number of different investigators in my lab, including Jeffrey Koshi, Darin Taverna, Matthew Dimmic, and Orkun Soyer, as well as a collaborator, Richard Neubig. Financial support was provided by NIH Grants GM08270 and LM0577, NSF equipment grant BIR9512955, and a grant from the University of Michigan Program in Bioinformatics.

REFERENCES

[1] Wako, H. & Blundell, T. (1994). Use of amino acid environment-dependent substitution tables and conformational propensities in structure prediction from aligned sequences of homologous proteins. I. Solvent accessibility classes. *J. Mol. Biol.* **238**:682-692.

[2] Wako, H. & Blundell, T. (1994). Use of amino acid environment-dependent substitution tables and conformational propensities in structure prediction from aligned sequences of homologous proteins. II. Secondary structures. *J. Mol. Biol.* **238**:693-708.

[3] Koshi, J. M., & Goldstein, R. A. (1995). Context-dependent optimal substitution matrices. *Protein Engineering* **8**:641-645.

[4] Goldman, N., Thorne, J. L., Jones, D. T. (1996). Using evolutionary trees in protein secondary structure prediction and other comparative sequence analyses. *J. Mol. Evol.* **263**:196-208.

[5] Koshi, J. M. & Goldstein, R. A. (1998). Models of natural mutations including site heterogeneity. *Proteins* **32**:289-295.

[6] Koshi, J. M. & Goldstein, R. A. (2001). Analyzing site heterogeneity during protein evolution. Pacific Symposium on Biocomputing. **6**:191-202.

[7] Dimmic, M. W., & Goldstein, R. A. (2000). Modeling evolution at the amino acid level using a general fitness model, in Pacific Symposium on Biocomputing 2000, (Altman, Dunker, Hunger, Lauderdale, and Klein, eds), World Scientific, Singapore, pps. 18-29.

[8] Soyer, O., Dimmic, M. W., Neubig, R. R., Goldstein, R. A., Modeling protein evolution with applications to the understanding of G-Protein Coupled Receptors. *Proteins*, submitted.

[9] Kidera, A., Konishi, Y., Oka, M., Ooi, T., Scheraga, H. A. (1985). Statistical analysis of the physical properties of the 20 naturally occurring amino acids. *J. Prot. Chem.* **4**:23-55.

[10] Hobohm, U. & Sander, C. (1994). Enlarged representative set of protein structures. *Protein Sci.* **3**:522-524.

[11] Sander, C. & Schneider, R. (1991). Database of homology-derived protein structures and the structural meaning of sequence alignment. *Proteins* **9**:56-68.

[12] Higgins, D. G., Bleasby, A. J., Fuchs, R. (1992). Clustal V: Improved software for multiple sequence alignment. *CABIOS* **8**:189-191.

[13] Kabsch, W. & Sander, C. (1983). Dictionary of protein secondary structure: Pattern recognition of hydrogen-bonded and geometrical features. *Biopoly.* **22**:2577-2637.

[14] Bernstein, F. C., Koetzle, T. F., Williams, G. J. B., Meyer, E. F. Jr., Brice, M. D., Rodgers, J. R., Kennard, O., Shimanouchi, T., Tasumi, M. (1997). The protein data bank: A computer-based archival file for macromolecular structures. *J. Mol. Biol.* **112**: 535-542.

[15] Horn, F., Weare, J., Beukers, M. W., Horsch, S., Bairoch, A., Chen, W., Edvardsen, O., Campagne, F., Vriend, G. (1998). GPCRDB: an information system for G protein coupled receptors. *Nucl. Acid Res.* **26**:275-279.

[16] Felsenstein, J. (1993). PHYLIP - Phylogeny Inference Package. *Cladistics* **5**:164-66.

[17] Yang, Z. (1994). Maximum Likelihood Phylogenetic Estimation from DNA Sequences with Variable Rates over Sites: Approximate Methods. *J. Molec. Evol.* **39**:306-314.

[18] Kawashima, S., Ogata, H., Kanehisa, M. (1999). AAindex: Amino acid index database. *Nucleic Acid Res.* **27**:368-369.

[19] Fraczkiewicz, R. & Braun, W. (1998). Exact and efficient analytical calculation of the accessible surface areas and their gradients for macromolecules. *J. Comp. Chem.* **19**:319-333.

[20] Koshi, J. M. & Goldstein, R. A. (1997). Mutation matrices: correlations and implications. *Proteins* **27**:336-344.

Pattern Recognition and Distributed Computing in Drug Design

W. Graham Richards

Department of Chemistry, University of Oxford, Central Chemistry Laboratory, South Parks Road, Oxford OX1 3QH, UK

E-Mail: graham.richards@chem.ox.ac.uk

Received: 30th April 2002 / Published: 15th May 2003

Abstract

Computational methods developed in the area of medical imaging can be adapted to find ligand binding sites on proteins. Once the binding site is specified, libraries of real or virtual molecules may be screened to seek out compounds which have very strong affinity. Massively distributed computing enables huge numbers of molecules to be screened.

These approaches will be illustrated by reference to a search for inhibitors of the binding of anthrax lethal factor to the protection antigen. With the site identified some 3.5 billion molecules were tested in 24 days using the power of 1.4 million personal computers running a screen saver. Over 300,000 hits were revealed with approximately 12,000 looking particularly promising.

Introduction

The origin of complexity in drug design is obvious: the sheer number of potential molecular structures with drug-like properties. The completion of the first stage of the human genome project revealed, somewhat surprisingly, that there are only a few tens of thousands of genes. This limits the number of protein targets which might be the starting point for drug design to at most a few hundred thousand. There are no such limitations on the molecular structures of drug candidates. Even if we demand a molecular weight range of between say 150 and 800 together with water-soluble compounds and synthesizability, only imagination limits the possibilities. There are billions of possibilities.

Two developments, one scientific and the other technological, have emerged in mitigation: pattern recognition techniques and massively distributed or grid computing.

Pattern Recognition

Although rarely invoked in the chemical literature, pattern recognition is a major topic in the field of engineering. Enormous effort has been expended in areas such as computer vision; medical imaging and perhaps above all in military applications. In a series of recent papers we have adapted some of the techniques to aspects of drug discovery.

The first problem which we have attempted to overcome using pattern recognition is that of the alignment of molecules which is the key prerequisite to inferring the nature of a binding site when we start without knowledge of the protein target structure at the atomic level. (1,2) Although there are a plethora of alignment techniques, almost all fail if one tries to superimpose one structurally optimally on top of a second structure when those two molecules are of very different sizes. Simple methods, which start by matching centres of mass, will overlay by placing the smaller molecule in the middle of the larger one. A technique from computer vision (2) which employs local structure analysis evades this trap. A sample example is shown in Figure 1.

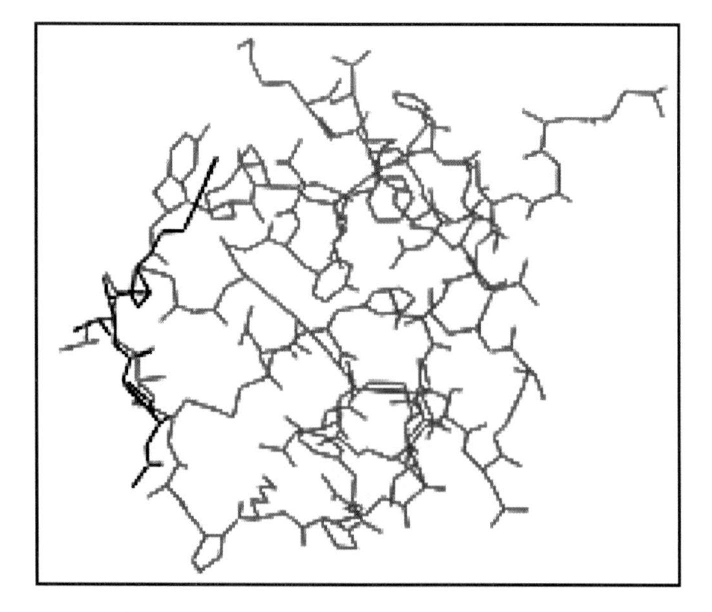

Figure 1. The computed alignment between DFKi (in black) and TOMI (in red) .

Here the larger molecule, a natural product turkey ovomucoid inhibitor is matched with the small molecule mimic DFKi which is a standard list introduced by Masek et al. (3). The small molecule is predicted to match exactly in the position where the surface active peptide fragment is situated and the matching only takes seconds on a personal computer.

More important still has been an adaptation of the multiscale approach from medical imaging (4) to find binding sites on proteins of known structure but unknown target binding site. As in medical imaging one does not want too much detail too soon or else one is overwhelmed by complexity. We need to do something very quickly and crudely but in an algorithmic stepwise manner more progressively to greater detail as we home in on the answer we are seeking. Techniques of this nature are for example used in the automatic scanning of mammograms. At the molecular level we can, in principle, find the binding site of a ligand on a protein by computing interaction energies between ligand and protein for all possible relative positions and orientations of the partners. This is the classic docking problem. Most available software essentially cheats by starting with the small molecule in roughly the correct position, perhaps already in a binding groove on the protein surface. Pattern recognition enables one to avoid this bias and still dock the pair in seconds by a hierarchical approach, illustrated in Figure 2.

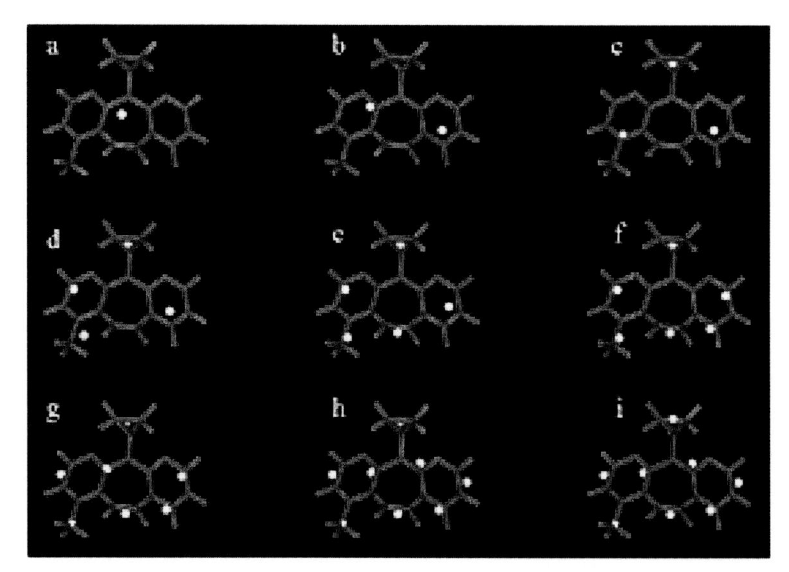

Figure 2. The hierarchy of models of nevirapine.

Firstly the small ligand is treated as a single point at the centre of mass with all the parameters for a molecular mechanics interaction energy collapsed and averaged on to that point. If we place limits on plausible interaction energy then the very rapid calculation of interaction between a point molecule and the protein will immediately restrict those parts of space where binding is not possible: not too far away to interact nor too close so as to have atom clashes. We then step up to a two-point model using the so-called k-means algorithm. To move from a one-cluster to a two-cluster or two-point model, we first find the atom furthest from the initial cluster and make this the temporary centre of the second point. All atoms closer to this centre than the original one are assigned to the second cluster and then the positions are iterated until self-consistent where each cluster centre is positional at the mean position of the atoms which belong to that cluster. This is then repeated in a stepwise fashion yielding one, two, three, four-centre models, but the interaction energy calculation is more restricted in space as the model grows in complexity. Our experience is that usually four steps are sufficient to locate the binding site to an accuracy comparable with the resolution of the crystal structure. This may then of course be optimized by employing more rigorous energy optimization procedures.

Once one has both a protein structure and a binding site, the essential feature of drug design is to try as many small molecules as possible in that binding site to see if they fit, preferably allowing flexibility in the protein and conformational freedom in the small molecule. One should then ideally compute relative binding free energies. One rapid manner of achieving these aims, albeit crudely, is to match patterns of pharmacophores. From the protein binding site structure one defines a set of binding points of obvious types: hydrogen bond donors and acceptors; lipophilic groups; changes; aromatic rings. All have positions in space. One then seeks complimentary binding contributors on the ligand for all possible shapes, allowing some leeway in distance constraints to permit a little macromolecular flexibility. When complimentary centres are found, and four such matches would seem to be ideal as this would incorporate stereochemistry in the ligand, a crude binding free energy may be computed.

The crude binding free energy will most easily be derived from a scoring function which adds up the number of specific interactions with a standard energy contribution for each (e.g. each hydrogen bond contributing 3.3 kcal mol^{-1}). The effect of the loss of conformational entropy on binding may be estimated by counting the number of rotatable bonds which are frozen and multiplying this by a factor. In addition a simple calculation using a molecular mechanics formula will give an idea of the van-der-Waals interaction energy and ligand torsional contribution. Thus matching pharmacophore patterns and then estimating energy contributions provides a measure of drug potential, although the utility will depend on the database of small molecules tried.

SMALL MOLECULE DATABASE

In our much publicised cancer drug project (www.chem.ox.ac.uk) a total of 3.5 billion small molecules have been screened as inhibitors of proteins, twelve of which are implicated as suitable targets for anti-cancer agents and also one as a site for blockers of anthrax which is discussed below.

To achieve such a large database, Davies (5) reviewed 1.4 million molecules which can be found in catalogues, and over one billion which have been made in published chemical libraries. These were then filtered to ensure that all had drug-like properties in terms of molecular weight and solubility (judged by having nitrogen or oxygen providing at least 20% of surface area). This filtering left 35 million molecules. For each of these 100 *de novo* structures were created by substitution of groups, such as $-CH_3$ for $-OH$ etc.

Although this is probably the biggest database of molecules ever screened, it must be emphasised that these lists have not been scrutinised for ease of synthesis, and one could envisage creating an even bigger and more useful database where the knowledge of organic and medicinal chemists could be incorporated. It is tempting to set up a website which is totally open where chemists the world over just contribute structures which they believe are synthesizable but with no limits on imagination. In that way one would hope to create a database with as wide a coverage of chemical space as is conceivable.

DISTRIBUTED COMPUTING

Even using fast pattern recognition techniques the screening of 3.5 billion molecules against protein targets is an enormous computational task. We have managed to achieve this using the methods of distributed computing in collaboration with United Devices Inc. (www.ud.com). The code to perform the calculations described above has been incorporated into a screensaver in a project started in April 2001. In a period of twelve months over 1.5 million personal computers have joined the project from over 200 countries, and contributed more than 100 thousand years of CPU time. It is currently (April 2002) possible to screen the database of 3.5 billion molecules in three weeks. (6)

The project is managed by a central server which dispatches work units of a number of small molecular structures and receives and validates the results. Part of the screensaver communicates with the server while the rest performs the calculations described above.

An Example: The Anthrax Project

The events following September 11, 2001 highlighted the dangers posed by the use of anthrax as a weapon of bioterror and accelerated scientific publication of relevant data. The essential molecular aspects of the action of anthrax are summarised in Figure 3.

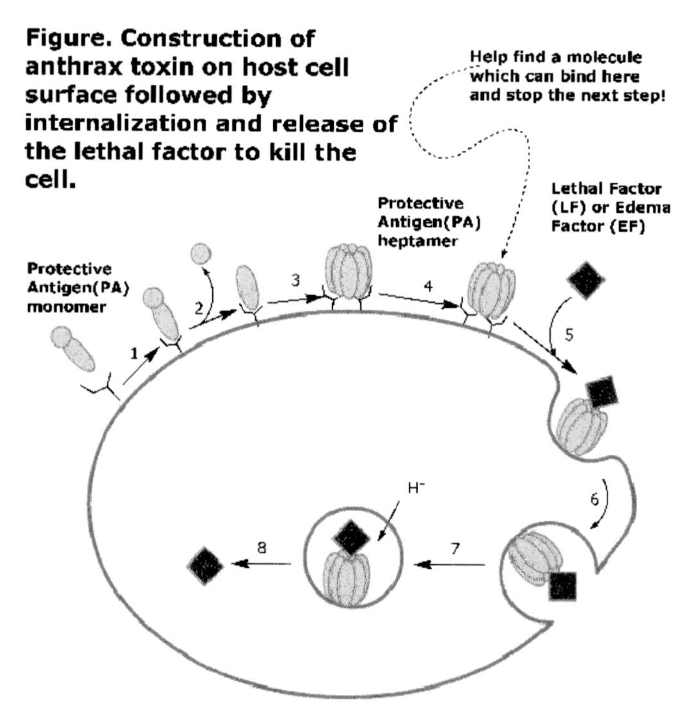

Figure 3. A representation of the mode of action of the anthrax toxin.

The toxin is comprised of three proteins. The so-called protective antigen (PA) forms a heptamer which facilitates the entry into the cell of the lethal factor (LF) and the edema factor (EF): individually the proteins are non-toxic. Work by Mourez et al (7) showed that a non-natural peptide where bound to a flexible polymeric backbone inhibits the binding of PA to LF/EF and protected rats against anthrax. Peptides which are active contained the short sequence YWWL which is not present in the anthrax proteins, suggesting that this hydrophobic peptide plays a role in binding to the PA heptamer so as to preclude the essential binding of LF/EF. One is thus in the situation described above where one has a huge protein of known structure and information about a significant ligand but no knowledge of its binding site.

Application of the pattern recognition software (8) highlights the binding site, which when once located proves to be at reasonably obvious and convincing position as illustrated in Figure 4, at the junction between proteins where LF/EF might be expected to bind but in precise atomic detail with very obvious hydrogen bonds as in Figure 5.

Armed with this information it was possible, using the distributed computing and screensaver, to try all the 3.5 billion molecules of the database. Using the then 1.4 million PCs the task was completed in 24 days, yielding some 300,000 crude hits of which about 12,000 look promising enough to merit further more refined scrutiny.

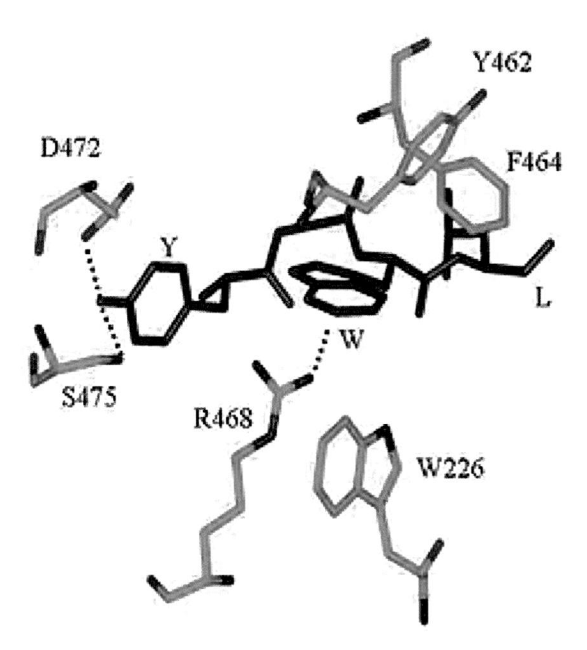

Figure 4. The heptamer of the anthrax protection antigen with the predicted bound tetrapeptide in blue.

CONCLUSION

The complexity of dealing with billions of potential drugs is eminently tractable if we combine the twin approaches of pattern recognition and distributed computing. Pattern recognition permits the specification of binding sites on proteins which is clearly going to become more and more important as structures flow from structural genomics. Pattern matching of pharmacophores for first-pass screening of massive databases is an ideal way to use the almost limitless power of distributed computing.

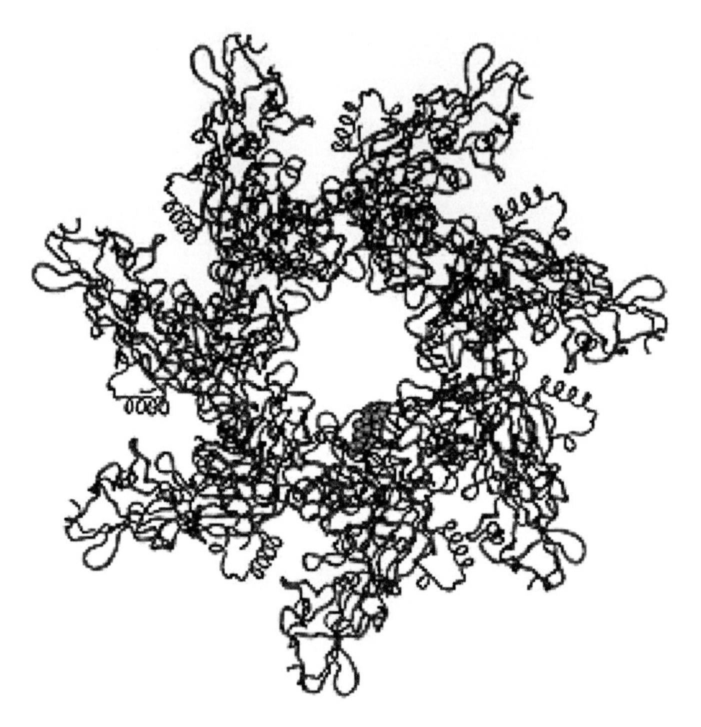

Figure 5. A detailed view of the tetrapeptide binding site of the anthrax toxin.

In the future much work of this nature is likely to be done in-house within the walls and indeed fine walls of pharmaceutical houses, but it would be a great pity if the open use of the web as employed in our cancer project were not to continue. Not only does this provide a huge, essentially free resource, the involvement of the general public as a means of increasing public understanding and indeed participation in science should not be underestimated.

Acknowledgements

This work has been supported in part by the National Foundation for Cancer Research and the ScreenSaver Projects received sponsorship from Intel Inc. and from Microsoft Corporation.

References and Notes

[1] Robinson, D. D., Lyne, P. D., Richards, W. G. (1999). Alignment of 3D-structures by the method of 2D-projections. *J. Chem. Inf. Comput. Sci.* **39**:594.

[2] Robinson, D. D., Lyne, P. D., Richards, W. G. (2000). Partial molecular alignment via local structure analysis. *J. Chem. Inf. Comput. Sci.* **40**:503.

[3] Masek, B. B., Marchant, A., Mathew, J. B. (1993). Molecular skins: a new concept for quantitative shape matching. *Proteins: Struct. Funct. Genet.* **17**:193.

[4] Glick, M., Robinson, D. D., Grant, G. H., Richards, W. G. (2002). Identification of ligand binding sites on proteins using a multiscale approach. *J. Am. Chem. Soc.* **124**:2337.

[5] Davies, E. K. submitted for publication.

[6] Davies, E. K., Glick, M., Harrison, K. N., Richards, W. G. Pattern recognition and massively distributed computing. *J. Comp. Chem.* in press.

[7] Mourez, M., Kane, R. S., Mogridge, J., Metallo, S., Deschatelets, P., Sellman, B. R., Whitesides, G. M., Collier, R.J. (2001). Designing a polyvalent inhibitor of anthrax. *Nature Biotechnol.* **19**:958.

[8] Glick, M., Grant, G. H., Richards, W. G. (2002). Pinpointing anthrax inhibitors. *Nature Biotechnol.* **20**:118

Molecular Informatics: Confronting Complexity, May 13th - 16th 2002, Bozen, Italy

Physicochemical Properties and the Discovery of Orally Active Drugs: Technical and People Issues

Christopher A. Lipinski

Pfizer Global Research and Development, Groton Laboratories, Connecticut, 06340, USA
E-Mail: christopher_a_lipinski@groton.pfizer.com

Received: 13th June 2002 / Published: 15th May 2003

Abstract

Poor aqueous solubility is the largest physicochemical problem hindering drug oral activity. Among combinatorial libraries, poor solubility is a frequently encountered problem but poor permeability is seldom a problem. The relative importance of poor solubility vs. poor permeability as a source of poor oral activity depends on the method of lead generation. Solubility or permeability problems are not purely a technical issue of assay design or computational prediction. People and organizational issues are extremely important. A computational ADME filter like the "rule of 5" (1) is most effective when used prior to the beginning of experimentation.

Introduction

Physicochemical property changes in recent drugs makes finding orally active drugs more difficult. Poor solubility will be viewed as the predominant problem if lead generation is heavily dependent on high throughput screening. Poor permeability will be viewed as the predominant problem if leads arise from structure based design. Adverse property changes can be managed through appropriate use of computational and experimental strategies. A computational filter for orally active drugs like the "rule of 5" is most effective when used prior to the beginning of experimentation because at this stage people issues are minimized. In my opinion, there is a hierarchy of properties that can be controlled by chemistry. Tight structure activity relationships (SAR) equate with good control. Properties important to oral activity like solubility and permeability do not show tight SAR and hence need early computational prediction and early experimental assays. Screening for poor aqueous solubility is important regardless of the type of chemistry. It is important for both heterocyclic and peptido-mimetic compounds. Medium to high throughput solubility assays, for example turbidimetric solubility assays, are only useful in early discovery. Traditional thermodynamic solubility assays are most appropriate to the discovery development interface when the crystalline state of drugs is well characterized. In contrast to the screening for poor aqueous solubility, the value of screening for poor permeability depends much more on the chemistry chemo-type. Experimental permeability screening is most valuable for conformationally flexible compounds particularly those containing multiple charged groups. By contrast permeability screening for heterocyclic compounds particularly those containing few rotatable bonds may not be very useful unless the permeability problem is related to a biological transporter. Heterocyclic compounds containing few rotatable bonds are the frequent products of combinatorial chemistry and computational predictors for permeability suggest that few compounds in combinatorial libraries will exhibit a permeability problem.

METHOD

There is a systematic method to understand the causes and potential solutions to problems of poor physicochemical properties that are associated with poor oral absorption. This method involves a historical and database analysis on how physicochemical properties have changed with time from the era where problems with poor oral absorption were not so pronounced. This chemo informatic database method is very analogous to the rationale often given for studying history. In the affairs of man it is necessary to understand the past (history) to avoid in the future repeating the errors of the past.

Two technologies in lead discovery have to a considerable extent dominated the scenario of drug lead generation. These technologies are high throughput screening (HTS) and combinatorial chemistry. It is very easy to track the onset of these technologies by performing a simple citation analysis. I searched SciFinder 2001® software from Chemical Abstracts using the text string “high throughput screening” and the text string “combinatorial chemistry”. Both searches gave essentially an identical profile with a rapid increase in literature citations starting in about 1995. The similarity in the citation profiles is very reasonable. HTS is the rapid screening of large numbers of compounds in a biological assay. The biological screening process requires large numbers of chemistry compounds to be assayed. Combinatorial chemistry, the automated generally robotic synthesis of large numbers of chemistry compounds provides the material to be screened in the HTS assays. It is common today to find statements in magazine articles similar to the following "HTS and combinatorial chemistry have not lived up to their promise". These statements are partly true but misleading because they fail to differentiate between the early and later stage of combinatorial chemistry and whether the problem is in the HTS process or the combinatorial chemistry screening file. In my opinion there is not a problem with HTS. The problem lies in the fact that the first fifty percent of the history of combinatorial chemistry was badly flawed from an oral drug delivery perspective. The valid technology of HTS could not easily yield drug-like (orally active) drug matter if the combinatorial chemistry compound starting points were badly flawed.

There are two factors responsible for the production of badly flawed combinatorial compounds up to about the 1997-8 time period. The earliest factor in a time sense leading to the production of badly flawed combinatorial compounds was the actual method of robotic chemistry synthesis. A new technology tends to adapt pre-existing technology. In the case of combinatorial chemistry the pre-existing technology was the Merrifield solid phase synthesis of peptides. This automated method of peptide synthesis was in place before the advent of combinatorial chemistry and automated synthesis equipment was commercially available. Peptide scaffolds are capable of presenting interesting chemistry functionality in various regions of space and so the earliest combinatorial libraries were constructed using peptide scaffolds. Initially the work focused on naturally occurring α-amino acids and later with non-natural amino acids. Early workers were fascinated with the possibility of discovering compounds with potent *in vitro* activity. This focus was completely understandable given the difficulty of discovering a drug lead with potent *in vitro* activity in the decade of the 1980's. Peptide scaffold based combinatorial libraries did indeed generate potent *in vitro* active compounds in the new HTS screens but it took a number of years to realize that these initial HTS hits were very difficult to convert into orally active compounds. Naturally occurring α-amino acid bonds are metabolically unstable so these early peptide based libraries had little or no *in-vivo* activity. Another problem that was initially not appreciated is that a compound with more than just a few amide bonds can be quite impermeable through the gastrointestinal wall. Hence many of these early peptide scaffold based combinatorial libraries were very poorly absorbed by the oral dosing route.

The second factor responsible for the production of badly flawed combinatorial compounds up to about the 1997-8 time period can be traced to the inappropriate implementation of the concept of maximum chemical diversity. In the concept of maximum chemical diversity one tries to synthesize compounds with interesting chemical functionality displayed in as many directions as possible in three dimensional space. The idea is to display chemistry functionality likely to be involved in target recognition in as many areas of chemistry space as possible. The greater the coverage of chemistry space with appropriate chemistry functionality the greater the likelihood of detecting activity in an HTS assay. Initially workers did not know how much chemistry functionality was necessary. It seemed likely that more was better. For example building a compound from four fragments gave a greater display of functionality than building a compound from three fragments. Also the theoretical number of combinatorial compounds that could be produced from four fragments was much larger than from three fragments. This was attractive because of the logic that screening greater numbers of compounds increased the probability of finding an active hit in an HTS assay. Hence many combinatorial libraries (collections of compounds) were synthesized from four fragments. Again early workers were fascinated with the possibility of discovering compounds with potent *in vitro* activity and HTS screening of these early tetramer libraries did indeed result in HTS hits with potent *in vitro* activity. It took a period of time before researchers discovered that these potent *in vitro* tetramer library hits were not producing orally active compounds on subsequent medicinal chemistry optimization. The problem was that the average tetramer combinatorial compound is very large with a molecular weight perhaps in the 650 Dalton range. Compounds in this molecular weight range tend to be both very impermeable through the gastrointestinal wall and very insoluble. Hence the phenomenon of potent *in vitro* activity but very poor or no *in-vivo* activity was observed.

Several more minor factors exacerbated the reliance of most pharmaceutical companies on these early flawed combinatorial libraries. The Pfizer Groton Connecticut laboratories began HTS in the late 1980's before the advent of combinatorial chemistry. Realizing the need for massive numbers of compounds for HTS screening Pfizer began a massive campaign to purchase available compounds from academic laboratories. This effort was well funded and very successful and largely completed by 1994. As a result, purchase of academic compounds was not a viable option by the time other pharmaceutical companies realized the need for acquisition of large numbers of compounds for HTS. Pfizer had quite literally cleaned out the world wide academic supply. A second factor exacerbating the reliance of most pharmaceutical companies on these early flawed combinatorial libraries was the unreliability of the only remaining alternative compound source to combinatorial chemistry that existed in the early 1990's. In 1991 the Soviet Union ceased to exist and quite rapidly large numbers of Soviet block chemists became unemployed and had to feed themselves and their families. A large synthetic capacity existed in the former Soviet Union and a demand for compound supply certainly existed in the largely western pharmaceutical companies. This supply and demand should in theory have resulted in a good match between supplier and customer. Unfortunately the quality control of these early Soviet block compounds was very poor. For example, in our Pfizer experience with these compounds we quite literally encountered the same compound sold to us with five different chemical structures. In our case this problem was not resolved until the compounds were delivered with appropriate spectral proof of identity. I believe our experience was probably shared by other companies. The result was that HTS screening results from these early Soviet block compounds was quite unreliable. In recent times the situation has completely changed. High quality compounds both combinatorial and non combinatorial accompanied by spectral documentation are now available from various vendors from the former Soviet block countries.

Aqueous solubility and permeability data must be provided to chemistry as early as possible to avoid oral absorption problems.

The 3D graph (Fig. 1) illustrates the three parameters under chemistry control that determine whether a compounds physicochemical profile is compatible with oral activity. The chemist has to synthesize a compound to achieve the appropriate combination of potency, solubility and permeability to move the compound into the region of space occupied by an orally active compound (above the solid surface). The points below the surface represent possible starting points in a lead optimization process. Usually the starting point is inferior in all three properties. Very frequently if only potency is improved it may be impossible to achieve oral activity (even with high potency) if solubility and permeability are very poor. The optimization of potency at the expense of poor solubility and / or poor permeability is a common occurrence Medicinal chemistry *in vitro* potency improvement usually does not improve a solubility or permeability problem in the lead starting point. In fact the usual pattern is that *in vitro* activity optimization results in an increase in both molecular weight and lipophilicity. Increases in these properties tend to correlate with increased poor aqueous solubility. In theory, extremely high potency will solve a permeability or solubility problem. In practice, it is quite difficult to get orally active drugs at doses below 0.1 mg/kg.

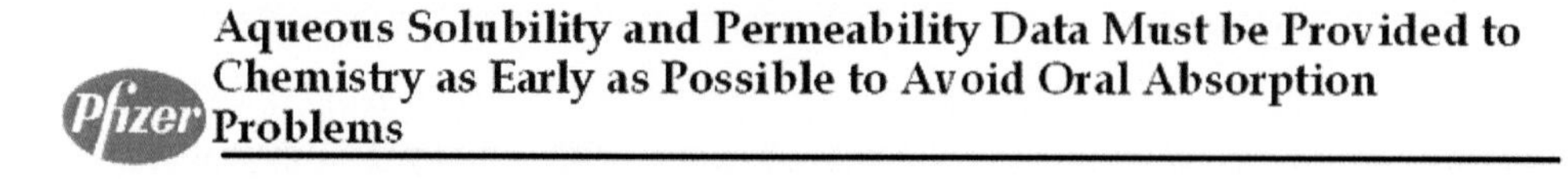

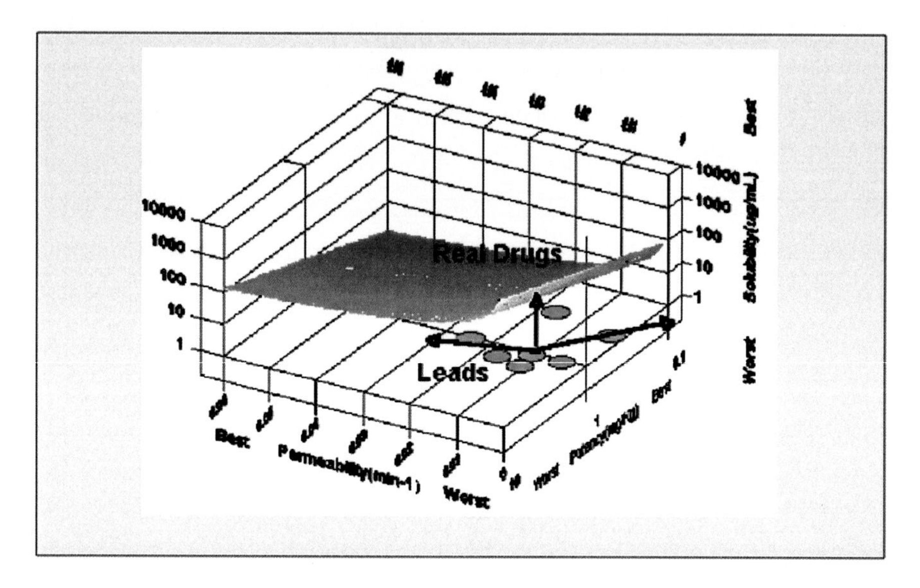

Figure 1.

The reason is that at very low doses a variety of metabolic processes can easily destroy the drug. At higher drug doses, these metabolic processes are saturated and less important. In my opinion, it is often easier to solve solubility problems than to solve problems in passive membrane permeability since the range in drug-like solubility is much greater than for permeability. For example, the FDA's proposed bio-equivalence classification system (BCS) classifies drugs into 4 classes depending on whether drugs have high or low permeability and high or low solubility. In the BCS, the range for permeability covers considerably less than three orders of magnitude while that for solubility covers a full six orders of magnitude. The best way to solve a permeability or solubility problem is with chemistry. The key to avoiding this problem is to provide the chemist with information on solubility and permeability at the same time as the potency information is received.

The General Pharmaceutics Laboratory in our development organization profiles all newly nominated clinical candidates. As part of the evaluation, a minimum absorbable dose (MAD) is calculated for oral dosage forms based on the expected clinical potency, the solubility and the permeability. This calculation serves to confirm that either the physicochemical properties of the candidate are easily within the acceptable range or that the properties lie within a difficult range that will require more than the average pharmaceutics manning to solve any difficulties.

We have adapted this calculation to create a simple bar graph (Fig. 2) that we distribute to our medicinal chemists. It answers the question of "how much solubility does the chemist need?" Presented in bar graph format the information is very readily understood by our chemists. Presented to our chemists in the original published equation format its impact on our chemists was poor. Bar graph (2) illustrates a people issue. It is intended for presentation to our medicinal chemists and uses information from a paper published by a Pfizer pharmaceutical sciences researcher on a minimum absorbable dose (MAD) calculation. It is very important to present information in a format readily grasped by the intended audience. Pharmaceutical scientists are very comfortable with information presented in an equation format. Synthetic organic chemists are uncomfortable with mathematic equations.

There is a simple reason for this. American Ph.D. granting programs require four semesters of calculus to obtain a Ph.D. in chemistry. However calculus is not needed to be a competent synthetic chemist. All that is really needed is the mathematical skill set that typically comes from a quality high school education. Synthetic chemists tend to forget those math skills that are not needed in their profession.

Minimum Acceptable Solubility in ug/mL Bars shows the minimum solubility for low, medium and high permeability (Ka) at a clinical dose. The middle 3 bars are for a 1 mg/Kg dose. With medium permeability you need 52 ug/mL solubility.

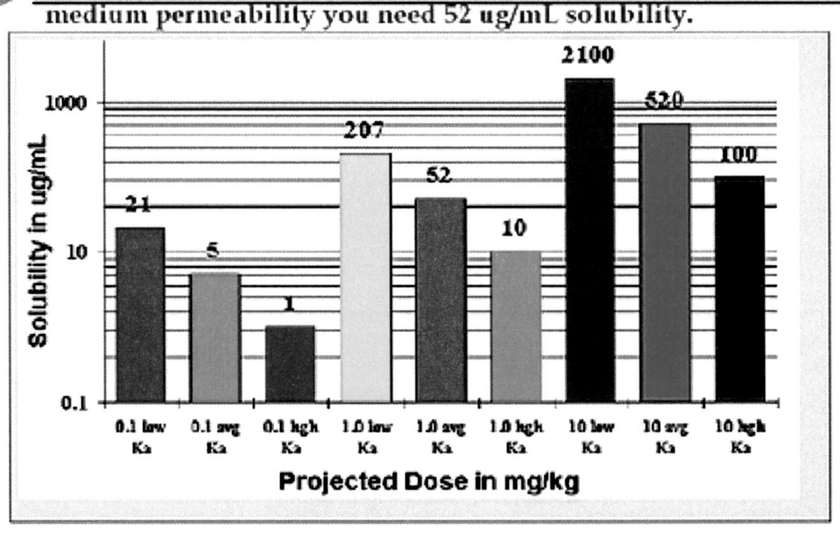

Figure 2.

By way of contrast synthetic chemists have very finely tuned pattern recognition skills with the ability to read a tremendous amount of information from a chemical structure depiction. This pattern recognition skill carries over to a graphical representation like a bar graph.

Minimum Acceptable Solubility for a drug can be calculated using an equation that takes into account the drug dose (potency), the solubility, the anticipated permeability and the intestinal fluid volume (assumed to be a constant). The usual solubility concentration units are µg/ml. For a molecular weight of 500 Daltons 5 µg/ml translates to a molar concentration of 10 µM. The acceptable solubility ranges are displayed in bar graph (2). Each set of three bars shows the minimum solubility for compounds with low, medium and high permeability (K_a) at an anticipated clinical dose. The middle set of three bars is for a 1 mg/kg dose. With medium permeability you need 52 µg/ml solubility. The three middle bars describe the most common clinical potency that we encounter; namely that of 1 mg/kg. If the permeability is in the middle range as for the average heterocyclic (the purple bar) then a thermodynamic solubility of about 50 µg/ml at pH 6.5 or 7 is required. If the permeability is low (as in a typical peptido-mimetic) the solubility should be about 200 µg/ml.

Leads at Pfizer and in the drug industry in general, now trend toward higher molecular weight and lipophilicity. Bar graph (3) shows the trend in molecular weight for compounds synthesized in our medicinal chemistry labs (shown in red) and compounds purchased from external commercial sources (shown in blue). In our Pfizer Groton laboratories we began HTS screening in 1989, and increased HTS through 1992. The percentage of compounds with a molecular weight over 500 (which we believe is undesirable) tracks exactly with the increased HTS screening. More and more of our leads were from HTS, these had poorer physicochemical profiles and when our medicinal chemists followed up these leads they made compounds with profiles like those of the leads or sometimes even worse than those of the leads. The trends in compounds made in our medicinal chemistry labs are not aberrant; they are completely logical (and predictable) in terms of medicinal chemistry principles and the information available to the chemists. For example, introducing a lipophilic moiety (e.g. a methyl) so as to fit into a receptor is one of the best ways to improve *in vitro* potency. This same change however, also increases lipophilicity. Compounds purchased from commercial sources (in blue) were intended for random HTS screening and show no upwards trend in high MWT. A bar graph with high lipophilicity instead of high molecular weight would look very similar.

Computationally comparing libraries allows one to deduce the differences between real drugs and those medicinal chemistry compounds which do not really possess drug-like properties. One can use the presence of an International Non-Proprietary Name (INN name) or a United States Adopted Name (USAN name) or marketed status as a marker for a compound with "drug-like" properties. Inn names and USAN names are assigned when a compound enters phase two clinical efficacy studies. Entry into clinical phase II efficacy study is a marker for drug-like properties.

Compounds that fail to survive the phase I human toleration studies or the pre-clinical stage do not receive an INN or USAN name. The compounds with severe oral absorption, toxicity and metabolism issues have been filtered out in a compound achieving phase II status. A compound entering into phase II is a real drug in the sense that except possibly for efficacy it has all the attributes of a real drug. Historically, of those drugs reaching phase II, ninety percent have been intended for oral administration.

Leads at Pfizer and in the drug industry in general, now trend toward higher MWT and lipophilicity

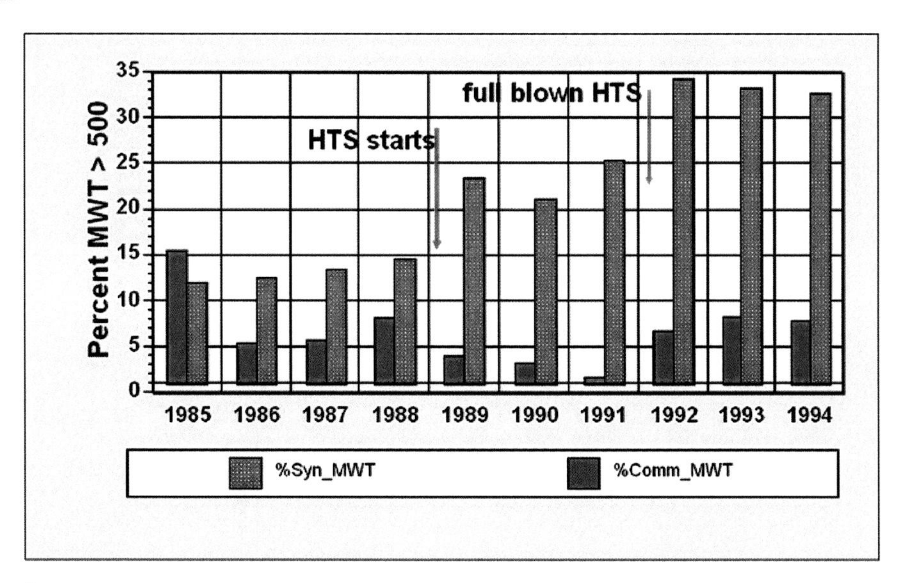

Figure 3.

So the presence of an INN or USAN name reliably identifies a set of orally active drugs with "real" drug like properties.

Drug-like as opposed to non drug-like physicochemical characteristics can now be defined by comparing drug-like with non drug-like data sets. The drug-like data set is a set of 7483 drugs which encompasses drugs with an INN/USAN name as well as drugs that were actually approved for marketing in at least one country. The library of 7483 INN/USAN and marketed drugs that was our benchmark is a significant fraction of all drugs that have reached phase II status. For example, there were about 9,500 USAN drugs listed in the most recent publication of the US Pharmacopeia.

The non drug-like data set is a set of 2679 drugs with the character that they represent a much earlier stage of drug discovery before any significant filtering for drug-like properties has occurred. I obtained this data set from the Derwent World Drug Index using the following procedure. I looked for drugs where the mechanism field contained the text "trial preparations". This procedure identified drugs intended for a medicinal therapeutic purpose. I excluded any drug with a Chemical Abstracts Service (CAS) registry number. This effectively made sure that the drug had only recently been abstracted into the Derwent World Drug Index (WDI) because I knew from experience that it typically took about two years for the CAS registry number to be included in the WDI. I also double checked that no compound in the non drug-like data set had an INN/USAN name. The compounds in the non drug-like data set were all abstracted in 1997, 1998 and 1999. This data set will off course contain some real drug-like compounds but it will also contain many more compounds that are only ligands for a biological target and that lack some or all of the attributes required for an orally active drug. This non drug-like data set represents the type of early discovery stage compound that one is likely to encounter prior to any filtering operation. Compounds similar to this data set are likely to be encountered in preliminary reports of biological activity at scientific meetings and to appear in the primary medicinal chemistry literature.

I have compared the distribution of the physicochemical properties for the drug-like compounds and the non drug-like compounds in figure 4. The reader can also think of these data sets as corresponding to the newer (non drug-like) and older (drug-like) compounds.

A convenient method of comparing the distribution of a property across non equally sized data sets is to compare Kaplan-Meier type survival curves. This graph shows the distribution of molecular (formula) weights of four classes of compounds. Shown in blue are drugs with International Non Proprietary Names INN) and United States Adopted Name (USAN) name. These are the drug-like compounds that have survived phase I with sufficient oral bioavailability and acceptable pharmacokinetic and pharmacodynamic parameters to reach phase II. Shown in green are New Chemical Entities (NCE). These are the drugs that actually reached market and are the compounds that are summarized in the "To Market - To Market" chapter in the back of the issues of "Annual Reports in Medicinal Chemistry".

By definition these are certainly real drugs. Shown in yellow are compounds appearing in the Derwent World Drug Index. This includes a very wide range of compounds. All have some sort of biological activity. Shown in red are the new drugs (the non drug-like) data set of drugs that were abstracted by Derwent in 1997, 1998 and 1999 from the medicinal chemistry journal and conference literature. The MWT corresponding to the 90th percentile and a decreasing probability of oral activity is marked by a horizontal line.

Newer drugs (red) are larger

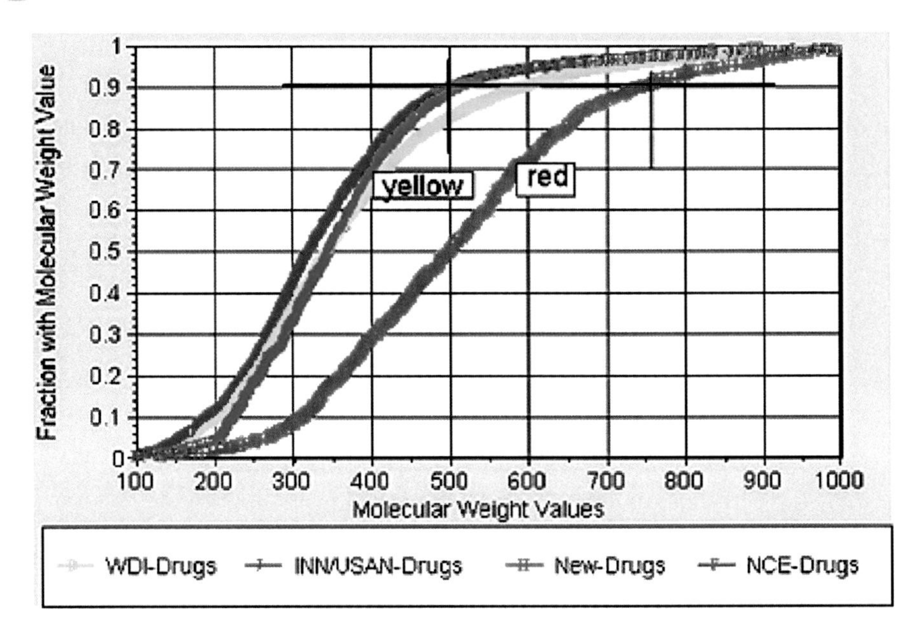

Figure 4.

When a curve is shifted towards the right it means that globally that data set has a higher distribution of the parameter. The red curve is distinctly shifted toward the right relative to all the other curves. This means that the new drugs overall have higher molecular weights than the real drug-like compounds. Newer drugs (non drug-like compounds) are larger in size than traditional, older real drugs. Figure 5 shows a set of Kaplan-Meier like curves for four physicochemical parameters in the same INN / USAN data set.

The idea is that the distribution of parameters for INN / USAN drugs can be used to define a property range where oral activity is increasingly difficult due to poor absorption or poor permeability. The distribution of four parameters for 7483 INN/USAN drugs define the ninety percent limits corresponding to properties unfavorable for oral drug absorption. The four properties were chosen based on extensive literature precedent. Too high a molecular weight was previously known to be linked to poor solubility and permeability. It was previously known that typically for a particular drug series there was an optimum lipophilicity for biological activity.

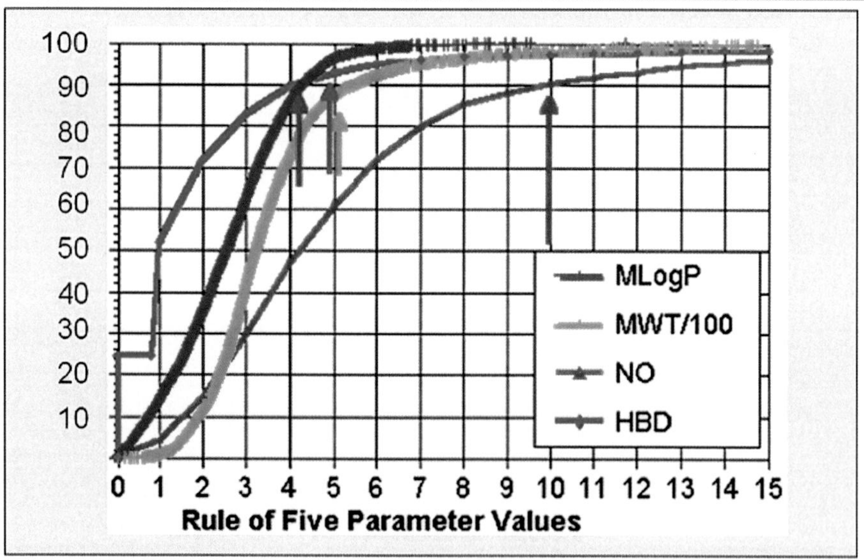

Figure 5.

Too little or too much lipophilicity was detrimental to biological activity. From work on peptides and peptide-mimetic compounds it was known that too many hydrogen bonding interactions between drug and water were detrimental to the ability of the drug to cross (permeate) the gastrointestinal wall. Permeation of the gastrointestinal wall is an absolute requirement for oral activity in a drug

All the curves exhibit a leveling as parameters reach unfavorable values for oral activity. The 90th percentile of each parameter is shown by the arrows. The colored lines show the distribution of: MLogP - lipophilicity (in blue) as measured by the Moriguchi Log P algorithm; MWT/100 - molecular weight (in light green), divided by 100 for plotting; OH+NH - the sum of OH plus NH (in red) as an index of H- bond donors; O+N - the sum of oxygen plus nitrogen (dark green) as an index of H-bond acceptors. There is a very clear similarity in the patterns of all four curves. For each parameter most of the values lie in the region below the ninety percent asymptote. It appears as if for real drugs that there are limiting values for all four of these parameters. The vast majority (ninety percent) of these real drug-like compounds do not exceed a particular parameter value.

RESULTS

This analysis led to a simple mnemonic which I called the "Rule of 5" because the parameter cutoff values all contained 5's. Numerically there actually are only four rules.

The "Rule of 5" states: Poor absorption or permeation are more likely when there are:

- More than 5 H-bond donors
- The MWT is over 500
- The CLog P is over 5 (or MLOGP is over 4.15)
- The sum of N's and O's is over 10
- Substrates for transporters and natural products are exceptions

Although this rule is very simple, it works remarkably well provided you understand its limitations. First, it only works because the physical property profile of medicinal compounds being currently made is quite far outside that of marketed drugs. Secondly, it doesn't work for compounds that are of natural product origin or have structural features originally derived from natural products, for example antibiotics, antifungals. The likely reason is the important roles of transporters in these classes. It also doesn't work well for certain therapeutic areas where many drugs are substrates for biological transporters. Anti-infective agents are a specific example of a therapeutic class where the "rule of 5" does not work well. Many anti-infective agents, e.g. the orally active cephalosporins are orally active because they are substrates of the PEPT-1 biological transporter. The affinity for the biological absorptive transporter allows the drug to bypass the physicochemical "rule of 5" limits for gastrointestinal wall permeability. Pfizer uses the "Rule of 5" in a variety of ways. For example it is used as an on-line alert at compound registration. It is used as a filter for high throughput screening (HTS) libraries. We do not screen libraries (collections) of compounds with significant non-compliance to the "rule of 5". We use the "rule of 5" as a filter for purchased compounds. We use it as a criterion for focused library synthesis and we use it as a guideline for quality clinical candidates. The latter use is not unique to Pfizer. In fact more and more there is recognition in the literature that the quality of the starting point in a chemistry optimization process is a good index of the final quality of the clinical candidate. We are now seeing analyses where researchers are tracking the relationship between the structure of a marketed drug to the structure of the starting point leading to the drug. The tight relationship between the starting point and the final drug is remarkable. A good starting point is likely to lead to a good drug. Conversely it is very difficult (but not impossible) to convert a poor starting point into a quality drug clinical candidate.

Considerable information relating to possible causes of poor solubility and poor permeability results from looking at how the properties of a drug company clinical candidates have changed with time (2). I am going to compare how important properties have changed with time for two very different drug organizations by comparing the properties of clinical candidates from the Pfizer Groton CT labs and the worldwide Merck organization.

Both organizations have been very successful in discovering drugs but they have done it in very different ways. For example one can plot the molecular weight (essentially a measure of size) for each early stage clinical candidate from Pfizer's Groton labs and fit the best straight line through the points. One see lots of scatter but the trend is clearly up. Over the years Pfizer Groton clinical candidates have gradually become larger. One can also discern the industry wide trend towards higher molecular weight in clinical candidates from Merck by analyzing the molecular weight trend with time for Merck advanced candidates (identified by MK numbers). Merck MK-numbers are issued in non sequential order and not all Merck MK compounds in the literature are candidates. For example, important biological standards may be assigned an MK-number. For this reason, the time scale for the analysis is the date of the earliest Merck patent corresponding to the MK-number candidate.

One can examines the trend of MWT as a function of time for each Merck candidate and fit the best trend line. Although there is considerable scatter there is clearly an upward trend in molecular weight with time. So just like Pfizer, Groton, Merck's clinical candidates have also gotten bigger with time. In a similar exercise one can plot the lipophilicity trend with time for Pfizer Groton clinical candidates. There is an upward trend with time for Pfizer Groton clinical candidates to become more lipophilic. It appears as if they are pushing right up to about a limit of 4-5 in logP. They don't go much higher because it really gets hard to get an orally active drugs when you exceed a value of 4-5 (the specific value of Log P varies a bit with the method of calculation). This just does not happen with clinical candidates from Merck. With time they absolutely do not become more lipophilic. So there has to be something very different about how Pfizer and Merck discover drugs. Not better or worse, just different.

Hydrogen bond acceptors are atoms in a drug that can accept an interaction with water through a hydrogen bond. Too few hydrogen bonds in many cases is not a good thing and too many hydrogen bonds is also not a good thing. More than about ten hydrogen bonds in a drug is not good because the drug will difficulty getting through the wall of the intestine into the blood. A drug given by mouth (an orally active drug) has to get from the inside of the intestinal tract through the intestinal wall to reach the blood stream. Certain kinds of atoms like Nitrogen (N) and Oxygen (O) in a drug accept these hydrogen bonds. So if you just count up the number of N's and O's in the drug molecular formula you get a simple (but still quite useful) measure of this hydrogen bond accepting property. There is a trend with time towards increasing number of hydrogen bond acceptors among Merck candidates. This trend is what one might expect given the strong focus in structure based drug design in recent years and on a type of chemistry called peptido mimetic like structures. This is the kind of drug discovery that Merck is famous for and very good at. A similar analysis for Pfizer Groton candidates would absolutely not show this upwards trend in hydrogen bond acceptors. Pfizer Groton does a lot of HTS whereas over the time period of this analysis Merck focused more heavily on all the various approaches to rational drug design other than HTS). One approach is not better than the other, just different. But the differences in approaches show up over time. So there must be something about HTS as opposed to non HTS drug discovery that leads to differences in trends towards increased hydrogen bonding functionality over time. So what do these differences between Pfizer, Groton and Merck and the differences in clinical candidate trends with time mean? Well with Merck the trend is towards larger size and more hydrogen bond interactions between the drug and water. Taken too far this translates to a problem in getting through the gastro intestinal tract wall. So an organization like Merck tends to worry about this property of poor permeability (problems getting through the gut wall). With Pfizer in Groton the trend is towards larger size and greater lipophilicity. Taken too far this translates to a problem in dissolving in the water inside the gastro intestinal tract. This is a problem of poor solubility and a drug has to be soluble to be orally active.

There is no free ride in drug research. Every discovery approach has a downside. Poor solubility and poor permeability are both bad. But they are not equally bad. It is much better to have poor solubility than poor permeability. The reason is that currently there are pharmaceutical sciences fixes for poor solubility. One would like to avoid them for all kinds of reasons but they do exist. By contrast, there is no pharmaceutical sciences formulation fix for poor permeability (except changing the chemistry structure as in a pro-drug) and there likely will not be for at least the next five years.

What are the reasons for the different physicochemical profiles in structure based as opposed to HTS based discovery approaches? In structure based approaches one is typically working on enzyme inhibitors or peptido-mimetics. Potency enhancement usually involves probing for at least three binding sites, e.g. in the P1, P1', P2 pocket. The binding pocket is often elongated. These considerations tend to lead towards larger size. Hydrogen bonding count tends to go up because one is often trying to satisfy multiple receptor hydrogen bonding interactions. Often the natural ligand is a peptide. There is not much selection pressure for log P to increase because a lot is known about the target. Lipophilicity does not play a role in discovering the lead series as it does in the HTS based discovery approach. Large size and increased H-bonding translates to a poorer permeability profile. HTS based approaches tend to bias towards larger size and higher lipophilicity because these are the parameters whose increase is globally associated in a medicinal chemistry sense with an improvement of *in vitro* activity. Larger size and higher lipophilicity translate to poorer aqueous solubility. Fortunately for HTS based approaches this bias can be corrected by appropriate filtering based on compound physicochemical properties.

Combinatorial libraries show a distinctive pattern with regards to permeability and solubility. Specifically, poor permeability is seldom encountered as a problem in a combinatorial library. As a result permeability profiles are not very dependent on chemistry synthesis protocol. Almost any protocol will result in compounds predicted to have an acceptable permeability profile. The reason relates to chemistry. It is actually quite difficult to construct a combinatorial library with permeability problems. It is difficult to make libraries with many hydrogen bond donors and acceptors in a combinatorial manner. To illustrate this point I analyzed a set of 47,680 combinatorial compounds from the same commercial source made according to 30 different synthesis protocols. There was little variation in average polar surface area (PSA) across the protocols and all but one of the thirty protocols gave an average PSA of less than 140 square Angstroms. A PSA of less than 140 square Angstroms suggests that passive trans membrane permeability (the most common drug permeability mechanism) will be quite acceptable.

Solubility profiles in contrast to permeability profiles can be very dependent on the chemistry synthesis protocol. The reason is that poor aqueous solubility is the major physicochemical problem found in combinatorial libraries. I calculated average aqueous solubility in μg/ml for the same data set of 47,680 combinatorial compounds from the same commercial source made according to 30 different synthesis protocols. The solubility program I used was a Pfizer developed model based on experimental data in our discovery turbidimetric solubility assay. The experimental assay outputs solubility in aqueous pH 7.0 phosphate buffer in the range < 5 to > 65 μg/ml using a turbidimetric end point. The computational model based on experimental solubility on 20,000 compounds bins solubility into three ranges; a low range of 10 mg/ml or less; a middle range of 15 to 60 μg/ml and a high range of 65 μg/ml or greater. About 80% of the experimental data used in the model building is evenly distributed between the low and high ranges. The model was tested against 10,000 experimental solubility measurements that were not part of the model building data set. About 80% of the test set data was predicted to lie in the low and high solubility bins and the accuracy of the prediction was 80%. The calculated solubility for the 47,680 combinatorial compounds differed markedly by synthesis protocol with significant populations of protocols at both extremes of solubility. This finding differs markedly from that of permeability. Solubility in combinatorial libraries depends very much on the synthesis protocol and it is very possible for some synthesis protocols to result in poorly aqueous soluble compounds. This and other examples I have investigated leads me to the conclusion that in general poor permeability is not a problem in combinatorial libraries but poor aqueous solubility is indeed a common problem.

The title of this article includes mention of "people issues". A specific example of a "people issue" is found in the area of experimental solubility profiling of combinatorial libraries. The question relevant to people issues is this. Is it possible to improve the solubility profile of a combinatorial library by incorporating experimental solubility feedback from early exemplars of a library? Understanding this question and how it relates to people issues requires an understanding of the stages involved in the experimental production of a combinatorial library. The experimental component of combinatorial library production typically involves two stages.

These are a protocol development stage followed by a production stage. In the protocol development stage the chemistry to translate the computational design into chemical reality is explored and optimized. Reaction conditions are explored and optimized. Steric and electronic boundaries for reaction components giving acceptable yields are defined and the reaction schemes are converted into formats suitable for robotic implementation. Invariably the protocol development step is the experimental rate determining step. Protocol development is much slower than library production. That is, it takes much longer to work out the chemistry than it does to actually make the compounds once the chemistry is worked out. Compounds first become available for experimental solubility testing in the protocol development stage. The critical issue is timing. How early can the compounds be obtained? The related people issue is this. How early must experimental data be obtained in order for people to change their behavior? Our experience has been that people (chemists) are not willing to change behavior if experimental feedback on solubility comes late in the rate determining step of combinatorial library construction. This is a people factor. When people have invested significant time in protocol development they are very unwilling to change their plans based on late developing experimental data. So finding out that there are likely severe solubility flaws in a library design late in the protocol development stage does not have value in terms of changing the library properties. The chemist has performed most of the work in the rate determining step and is unwilling to stop or radically change chemistry late in the process. The library goes into production regardless of the solubility profile of the exemplars if the feedback occurs late in protocol development. This problem is not easily solved. It is very difficult to obtain exemplars which span the chemical space of the protocol design at an early stage. How early would exemplars have to be obtained so that people factors would permit the experimental solubility data to make a difference? My guess is that it would have to occur in the first 10-15% of the protocol development stage to make a difference. The people factors are very strong. Once chemists strongly commit themselves to protocol development they are very reluctant to stop or even to make very radical changes.

The details of the causes of poor aqueous solubility are important in terms of understanding possible solutions. In a simplistic sense poor aqueous solubility can be thought of as arising from some combination of two distinctly different causes. The balance of these causes differs from compound to compound. At one extreme one cause of poor aqueous solubility lies in what can be termed the cavity making problem. For a compound to dissolve, a cavity (a hole) has to be made in water. Strong hydrogen bonds must be disrupted to make the cavity. This costs a lot in terms of energy. Some of this energy can be regained if the drug forms favorable interactions with water once it is placed in the hole. However if the compound is very lipophilic few favorable interaction will be formed with water. Hence a large lipophilic compound will be very insoluble in water. A large lipophilic compound requires a large cavity and forming the large cavity costs a lot energetically in terms of many broken hydrogen bonds and little of the energy cost will be reclaimed because there will be few favorable interactions between the lipophilic compound and water. This extreme of solubility is relatively easy to predict computationally. For example 75% of compounds whose lipophilicity exceeds the "rule of 5" limit of logP = 5 have poor aqueous solubility of less than 20 μg/ml in our discovery turbidimetric solubility assay.

At the other extreme one cause of poor aqueous solubility lies in what can be termed the crystal packing problem. A crystalline compound must be liberated from its crystal lattice before it can dissolve. The strongest crystal lattice interactions arise from intermolecular hydrogen bond interactions and packing interactions between the compound and its neighbors in the adjoining unit cells. These interactions can be visualized in the crystal packing diagram that can be generated from a single crystal x-ray. Melting point is the single simple property that is most useful in terms of characterizing crystal packing interactions. A high melting point is indicative of strong intermolecular crystal packing interactions and a high melting point compound is likely to have poor aqueous solubility. A rule of thumb is that a hundred degree increase in melting point decreases aqueous solubility by a factor of ten. This rule of thumb only holds for neutral compounds. The solubility of organic compound salts is not predicted by melting point. Unfortunately for the prediction of aqueous solubility, melting point data is completely absent from combinatorial libraries. Combinatorial compounds purified by automated methods are not subjected to crystallization and are usually isolated in amorphous form. No melting point data exists for these compounds. Clearly it would be very advantageous to have some type of computational prediction method for melting point as an indication of compounds likely to be insoluble because of crystal packing interactions. Unfortunately, reliable methods to predict melting point do not currently exist. Among compounds intended as drugs poor solubility due to strong crystal packing is common. For example in our extensive turbidimetric solubility testing over one half of compounds found to be experimentally insoluble were not excessively large or lipophilic. The prediction of poor solubility due to crystal packing remains a major unmet computational need in drug discovery because of the relevance to poor aqueous solubility.

CONCLUSION

Oral absorption depends on adequate solubility and intestinal permeability. A compound is insoluble because it is either too lipophilic or the inter molecular crystal packing forces for the compound are too strong. Globally, in the current era, poor aqueous solubility is the single largest physicochemical problem hindering drug oral activity. Among combinatorial libraries, poor solubility is a frequently encountered problem but poor permeability is seldom a problem. The relative importance of poor solubility vs. poor permeability as a source of poor oral activity is very dependent on the method by which leads are generated as can be seen by an examination of the time dependent trends in Merck vs. Pfizer, Groton clinical candidates. Dealing with solubility or permeability problems in an early discovery setting is not purely a technical issue of assay design or computational prediction. People and organizational issues are extremely important. Assay or computational results must be communicated to medicinal chemists in a manner that allows chemists to decide how to modify chemical structure. Communication with chemists is best when it takes advantage of the chemists' superb pattern recognition skills and is least effective when presented in an equation format or in terms that cannot be equated with chemical structure.

REFERENCES AND NOTES

[1] Lipinski, C. A., Lombardo, F., Dominy, B. W., Feeney, P. J. (1997). Experimental and computational approaches to estimate solubility and permeability in drug discovery and development settings. *Adv. Drug. Del. Rev.* **23**:3-25.

[2] Lipinski, C. A. (2000). Drug-like properties and the causes of poor solubility and poor permeability. *J. Pharm. Tox. Meth.* **44**:235-249.

Molecular Informatics: Confronting Complexity, May 13th - 16th 2002, Bozen, Italy

The Molecular Feature Miner MOLFEA

Christoph Helma, Stefan Kramer and Luc De Raedt

Albert-Ludwigs-Universität Freiburg, Institut für Informatik, Georges-Köhler-Allee, Geb. 079, D-79110 Freiburg im Breisgau, Germany
E-Mail: {helma, skramer, deraedt}@informatik.uni-freiburg.de

Received: 26th June 2002 / Published: 15th May 2003

Abstract

Inductive databases are a new generation of databases, that are capable of dealing with data but also with patterns or regularities within the data. A user can generate, manipulate and search for patterns of interest using an inductive query language. Data mining then becomes an interactive querying process.

The inductive database framework is especially interesting for bio- and chemoinformatics, because of the large and complex databases that exist in these domains, and the lack of methods to gain scientific knowledge from them. In this article we present an example for inductive databases: Molfea is the Molecular Feature Miner that mines for linear fragments in the 2D-structure of chemical compounds. In the methodological part we will explain the inner working of the Molfea algorithm, using a simple example. In the second part we will present applications to the NCI DTP AIDS Antiviral Screen database and several benchmark Structure-Activity Relationship problems in toxicology.

Introduction

The automation of experimental techniques in biology and chemistry has led to an enormous growth of biochemical databases. But the generation of data is only the first step towards a better understanding of the underlying biochemical mechanisms and processes. The second step – where computer science plays a central role – is the analysis of the data in order to find patterns and regularities of scientific interest. In a third step, these patterns have to be interpreted and related to current knowledge, in order to obtain new hypothesis and scientific insights. The whole process of identifying valid, novel, potentially useful and ultimately understandable patterns and models is called *Knowledge Discovery* (1).

Data Mining is a step in the Knowledge Discovery process, that consists of the application of statistics, machine learning and database techniques to the data. During the last few years, it became obvious, that a tight integration of advanced database technologies and data mining techniques would be very desirable. One of the most interesting proposals in this respect are *Inductive Databases* (2, 3, 4, 5). Inductive databases tightly integrate data and patterns, i.e. generalizations or regularities within the data, in a database. They provide also a query language and an inductive database management system that supports the querying of both patterns and data.

The query language allows the user to specify the patterns that are of interest (using a number of constraints e.g. on frequency, generality, syntax, etc., cf. below), the inductive database management system searches efficiently for the patterns that satisfy these constraints.

The inductive database framework is extremely attractive for bio- and cheminformatics because it provides a tool to support scientists in each of the three steps sketched above. Our favourite view of an inductive database user is a scientist who queries interactively an inductive database (possibly with the help of a graphical interface), who inspects the resulting patterns, obtains new ideas, and reformulates the original query until a new scientific insight is obtained.

In this paper, we present a domain specific inductive database called MOLFEA (Molecular Feature Miner). MOLFEA is an instance of the general *Inductive Database Framework*. Another instance is e.g. PROFEA, the Protein Feature Miner, an inductive database for the analysis of the secondary structure of proteins (6). In the next section we will explain the basic concepts of the MOLFEA algorithm using examples and analogies, readers who are interested in a formal presentation are referred to the original literature (7, 8). In the third section we will present some applications of MOLFEA to biomedical and toxicological databases. Finally we will discuss related work, limitations and future extensions in the last section.

METHODS

MOLFEA mines databases with chemical structures for fragments, i.e. linear sequences of atoms and bonds, that fulfil user defined criteria. Within MOLFEA, the user can specify the fragments of interest using simple but powerful *primitives*. Primitives may require e.g. that fragments have a minimum (resp. maximum) frequency on a set of compounds, or that they contain a given subfragment.

A query might request, for example, all fragments that are present in at least 90% of the active molecules but in less than 5% of the inactives. MOLFEA efficiently computes the solutions to these inductive queries using the level wise version space algorithm (8).

MOLECULAR FRAGMENTS

Fragments are linear sequences of non-hydrogen atoms and bonds that are present in molecules. `N-C-C-O`, for instance, is a fragment meaning: "a nitrogen connected with a single bond to a carbon connected to a carbon connected to an oxygen". A molecular fragment *f matches* an example compound *e* if and only if *f* is a substructure of *e*. For instance, fragment `N-C-C-C` matches the first example compound in the dataset *A* of Fig. 1.

In computer science terms, fragments are strings over an alphabet consisting of elements and bond types. The language of molecular fragments M has some interesting properties, which can be used to develop efficient algorithms:

- **Generality:** One fragment *g is more general* than a fragment *s* (Notation: $g \leq s$) if *g* is a substructure of *s* (e.g. `C-O` is more general than `N-C-C-O`). This has the consequence that *g* matches whenever *s* does.
- **Symmetry:** Two syntactically different fragments are equivalent, when they are a reversal of one another (e.g. `C-C-O` and `O-C-C` denote the same substructure).
- **Summary:** $g \leq s$ if and only if *g* is a subsequence of *s* or *g* is a subsequence of the reversal of *s* (e.g. `C-O` ≤ `C-C-O` and `O-C` ≤ `C-C-O`).

CONSTRAINTS ON FRAGMENTS

The fragments of interest can be specified by declaring constraints. Using the example datasets from Figure 1, it is e.g. possible to ask for fragments, that are present in at least two molecules from *A* and in not more than one compound from *B*. In more formal terms, one would formulate a query:

$$freq\ (f,\ A) \geq 2 \wedge freq\ (f,\ B) \leq 1$$

The whole query consists of a conjunction of primitive constraints $c_1 \wedge \ldots \wedge c_n$. Presently the following primitive constraints c_i are implemented in MOLFEA:

- $f \geq p$, $p \geq f$, $\neg (f \geq p)$ and $\neg (p \geq f)$: where f is the unknown target fragment and *p* is a predefined fragment; this type of primitive constraint denotes that *f* should (not) be more specific (general) than the specified fragment *p*; e.g. the constraint $f \geq$ `C-O` specifies that *f* should be more specific than `C-O`, i.e. that *f* should contain `C-O` as a subsequence;

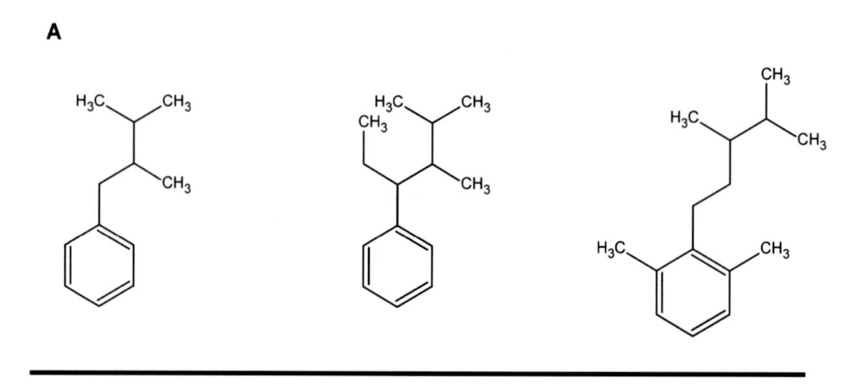

Figure 1. Example datasets A and B for the query $freq(f, A) \geq 2 \wedge freq(f, B) \leq 1$.

- $freq(f, D)$ denotes the frequency of a fragment f on a set of molecules D; the frequency of a fragment f in a database D is defined as the number of molecules in D that f matches;

- $freq(f, A) \geq t$, $freq(f, B) \leq t$ where t is a positive number and A and B are sets of molecules; this constraint denotes that the frequency of f on the dataset A or B should be larger than (resp. smaller than) or equal to t; e.g. the constraint $freq(f, A) \geq 2$ denotes that the target fragments f should match at least 2 molecules in the A set of active molecules. The first type of primitive is called a maximum frequency constraint, the second one a minimum frequency constraint.

These primitive constraints can be combined conjunctively in order to specify the fragments of interest. Note that the conjunction may specify constraints with respect to any number of datasets, e.g. imposing a minimum frequency on a set of active molecules, and a maximum one on a set of inactive ones. E.g. the following constraint:

$$(\texttt{C-O} \leq f) \wedge \neg (f \leq \texttt{N-C-C-C-O}) \wedge freq\ (f, X) \geq 200 \wedge freq(f, Y) \leq 10$$

queries for all fragments that include the sequence `C-O`, are not a subsequence of `N-C-C-C-O`, match more than 200 molecules in X and less than 10 molecules in Y.

SOLVING CONSTRAINTS: THE MOLECULAR FEATURE MINER MOLFEA

In this section we will demonstrate, how MOLFEA solves queries for fragments efficiently. A naive approach would possibly generate all fragments, that are present in the dataset and check, which of the fragments fulfil the given criteria. This method is, of course, computationally very expensive and infeasible for large datasets. But there are more efficient ways to accomplish the same goal.

We will start with a simple minimum frequency constraint $freq(f, D) \geq t$. This constraint has the important property of *anti monotonicity*. To illustrate anti-monotonicity let us assume, we have two fragments *g* and *s* and we know that:

- *g* is more general than *s* (i.e. $g \leq s$; e.g. *g*: `C-O`, *s*: `C-O-S`), and that
- *s* is a solution to our constraint (i.e. *s* matches at least *t* times in *D*)

Then the anti-monotonicity allows us to conclude, that *g* is also a solution. According the definition of generality, general fragments match whenever the specific ones do. The general fragment *g* must be therefore at least as frequent as the specific fragment *s*. Anti-monotonicity is also the reason, why we do not have to determine all solutions for a query. For anti-monotonic constraints, it is sufficient to know the set of the most specific fragments *S*, all fragments that are more general than an element in *S*, will also fulfil the constraint.

Maximum frequency constraints $freq(f, D) \leq t$, in contrast are monotonic. If *g* is more general than *s* and *g* is a solution, we know, that *s* must be a solution, because whenever *g* does not match, *s* will also not match. In this case we have to determine the most general set of fragments *G*, to determine all solutions.

The concept of determining borders, that completely characterize the set of solutions, is a well known idea in Machine Learning and Data Mining (9, 10, 11). It is especially useful, when we want to solve conjunctive queries, consisting of several primitive constraints c_i. In this case we can take advantage of their independency.

$$sol(c_1 \wedge \ldots \wedge c_n) = sol(c_1) \wedge \ldots \wedge sol(c_n)$$

So, we can find the overall solutions by taking the intersection of the primitive ones. In practice, we determine S_1 and G_1 for the first primitive constraint, and update S_i and G_i sequentially for each constraint, depending on the monotonicity of the primitive.

The basic anti-monotonic constraints in the MOLFEA framework are presently: $(f \leq p)$, $freq(f, D) \geq m$, the basic monotonic ones are $(p \leq f)$, $freq(f, D) \leq m$. Furthermore the negation of a monotonic constraint is anti-monotonic and vice versa.

So far we have not yet explained, how to find S or G for the primitive constraints. We will use the datasets A and B from Figure 1 as an example. If we are interested in finding all fragments, that occur at least twice in compounds A but not more than once in B, we can formulate the query

$$freq(f, A) \geq 2 \wedge freq(f, B) \leq 1$$

We will use the MOLFEA output in Figures 2 and 3 to illustrate the following procedure. Let's start with the first primitive constraint $freq(f, A) \geq 2$. The MOLFEA algorithm uses the set of the simplest possible fragments, the elements, as a starting point. These are the candidates for the first level. The next step is to eliminate those, that are too infrequent (i.e. $freq(f, A) < 2$). The remaining ones (`C`, `N`, `O` and aromatic carbon) fulfil our constraint, but they are not the most specific solution (i.e. there are longer fragments, containing the same elements, that are also frequent). So we have to generate more specific (i.e. longer) fragments for the next level.

Considering the generality relationship, we can take an important shortcut: It is not necessary to elongate the frequent fragments with all possible elements, but we have to consider only the frequent fragments {`C,N,O,C`}. Chemistry is not considered in this step, because "wrong" fragments are removed in the next elimination step. Again we check for frequencies, eliminate infrequent fragments and generate candidates for the next level by combining frequent fragments of the present level, under consideration of the symmetry relationship.

This process is repeated until no more specific fragments can be generated. The union of the most specific fragments is the new S_1 set {`C-C-C-O, N-C-C-O, C-C-C-C-C-C-C-C-C, N-C-C-C-C-C-C-C-C`}, the smallest (i.e. most general) fragments, that fulfil the constraint are the new G_1 set {`C, N, O, C`}.

The solution for the first primitive constraint G_1 and S_1 sets are the input for the algorithm that solves the second constraint $freq(f, B) \leq 1$. In this case, we have to remove those fragments that are too frequent in the dataset B. In other words, we have to update G to remove fragments that match more than one compound in the dataset B. Figure 3 shows the MOLFEA output for this toy example. We end up with a final G set {`C-C, C-O, C-C-C-C`} and S set {`C-C-C-O, N-C-C-O, C-C-C-C-C-C-C-C-C, N-C-C-C-C-C-C-C-C`}.

LEVEL 1:

Candidates: [Li], [Be], B, C, N, O, F, [Na], [Mg], [Al], [Si], P, S, Cl, [K], [Ca], [Sc], [Ti], [V], [Cr], [Mn], [Fe], [Co], [Ni], [Cu], [Zn], [Ga], [Ge], [As], [Se], Br, [Rb], [Sr], [Y], [Zr], [Nb], [Mo], [Tc], [Ru], [Rh], [Pd], [Ag], [Cd], [In], [Sn], [Sb], [Te], I. [Ca], [Ba], [Lu], [Hf], [Ta], [W], [Re], [Os], [Ir], [Pt], [Au], [Hg], [Tl], [Pb], [Bi], [Po], [At], [Rn], [Fr], [Ra], [Lr], c, n, s, o, p (78)

Frequent: C, N, O, c (4)

==

LEVEL 2:

Candidates: C-C, C-N, C-O, C-c, C=C, C=N, C=O, C=c, C#C, C#N, C#O, C#c, N-N, N-O, N-c, N=N, N=O, N=c, N#N, N#O, N#c, O-O, O-c, O=O, O=c, O#O, O#c, c-c, c=c, c#c (30)

Frequent: C-C, C-N, C-O, C-c, c-c (5)

==

LEVEL 3:

Candidates: C-C-C, C-C-N, C-C-O, C-C-c, N-C-N, N-C-O, N-C-c, C-N-C, O-C-O, O-C-c, C-O-C, C-c-c, c-C-c, C-c-C, c-c-c (15)

Frequent: C-C-C, C-C-N, C-C-O, C-C-c, C-c-c, c-c-c (6)

==

LEVEL 4:

Candidates: C-C-C-C, C-C-C-N, C-C-C-O, C-C-C-c, N-C-C-N, N-C-C-O, N-C-C-c, O-C-C-O, O-C-C-c, C-C-c-c, c-C-C-c, C-c-c-c, C-c-c-C, c-c-c-c (14)

Frequent: C-C-C-O, C-C-C-c, N-C-C-O, N-C-C-c, C-C-c-c, C-c-c-c, c-c-c-c (7)

==

LEVEL 5:

Candidates: O-C-C-C-O, O-C-C-C-c, C-C-C-c-c, c-C-C-C-c, N-C-C-c-c, C-C-c-c-c, C-c-c-c-c, C-c-c-c-C, c-c-c-c-c (9)

Frequent: C-C-C-c-c, N-C-C-c-c, C-C-c-c-c, C-c-c-c-c, c-c-c-c-c (5)

==

LEVEL 6:

Candidates: C-C-C-c-c-c, N-C-C-c-c-c, C-C-c-c-c-c, C-c-c-c-c-c, C-c-c-c-c-C, c-c-c-c-c-c (6)

Frequent: C-C-C-c-c-c, N-C-C-c-c-c, C-C-c-c-c-c, C-c-c-c-c-c, c-c-c-c-c-c (5)

==

LEVEL 7:

Candidates: C-C-C-c-c-c-c, N-C-C-c-c-c-c, C-C-c-c-c-c-c, C-c-c-c-c-c-c, C-c-c-c-c-c-C, c-c-c-c-c-c-c (6)

Frequent: C-C-C-c-c-c-c, N-C-C-c-c-c-c, C-C-c-c-c-c-c, C-c-c-c-c-c-c (4)

==

LEVEL 8:

Candidates: C-C-C-c-c-c-c-c, N-C-C-c-c-c-c-c, C-C-c-c-c-c-c-c. C-c-c-c-c-c-c-C (4)

Frequent: C-C-C-c-c-c-c-c, N-C-C-c-c-c-c-c, C-C-c-c-c-c-c-c (3)

==

LEVEL 9:

Candidates: C-C-C-c-c-c-c-c-c, N-C-C-c-c-c-c-c-c (2)

Frequent: C-C-C-c-c-c-c-c-c, N-C-C-c-c-c-c-c-c (2)

==

LEVEL 10:

Candidates: (0)

Frequency (0)

G: C, N, O, c (4)

S: C-C-C-O, N-C-C-O, C-C-C-c-c-c-c-c-c, N-C-C-c-c-c-c-c-c (4)

Figure 2. MOLFEA output for the first constraint of the query $freq(f, A) \geq 2 \wedge freq(f, B) \leq 1$.

The complete solutions for the example query is defined by these boarders. This means, that all subfragments of fragments in S that contain one of the fragments in G, are part of the solution *sol* = {`C-C, C-C-C, C-C-C-O, C-O, C-C-O,` ...}. Thus *S* and *G* together compactly represent the set of all solutions.

Results and Discussion

In this section, we briefly summarize our experiments with the DTP AIDS Antiviral Screen (http://dtp.nci.nih.gov) dataset and with toxicological *Structure-Activity Relationships* (SARs) using MOLFEA-generated features.

```
LEVEL 1:
Candidates:     C, N, O, c (4)
Frequent:       C, N, O, c (4)
Infrequent:     (0)
==
LEVEL 2:
Candidates:     C-C, C-N, C-O, C-c, C=C, C=N, C=O, C=c, C#C, C#N, C#O, C#c, N-N,
                N-O, N-c, N=N, N=O, N=c, N#N, N#O, N#c, O-O, O-c, O=O, O=c, O#O,
                O#c, c-c, c=c, c#c (30)
Frequent:       C-N, C-c, c-c (3)
Infrequent:     C-C, C-O (2)
==
LEVEL 3:
Candidates:     N-C-N, N-C-c, C-N-C, C-c-c, c-C-c, C-c-C, c-c-c (7)
Frequent:       C-c-c, c-c-c (2)
Infrequent:     (0)
==
LEVEL 4:
Candidates:     C-c-c-c, C-c-c-C, c-c-c-c (3)
Frequent:       c-c-c-c (1)
Infrequent:     C-c-c-c (1)
==
LEVEL 5:
Candidates:     c-c-c-c-c (1)
Frequent:       c-c-c-c-c (1)
Infrequent:     ( 0 )
==
LEVEL 6:
Candidates:     c-c-c-c-c-c (1)
Frequent:       c-c-c-c-c-c (1)
Infrequent:     (0)
==
LEVEL 7:
Candidates:     c-c-c-c-c-c-c (1)
Frequent:       (0)
Infrequent:     (0)

G: C-C, C-O, C-c-c-c (3)
S: C-C-C-O, N-C-C-O, C-C-C-c-c-c-c-c-c-c, N-C-C-c-c-c-c-c-c (4)
```

Figure 3. MOLFEA output for the second constraint of the query $freq(f, A) \geq 2 \wedge freq(f, B) \leq 1$. In dealing with this constraint MOLFEA starts from the results in Figure 2.

NCI DTP AIDS ANTIVIRAL SCREEN DATABASE

The NCI DTP AIDS Antiviral Screen program has checked more than 40,000 compounds for evidence of anti-HIV activity. The screen utilizes a soluble formazan assay to measure protection of human CEM cells from HIV-1 infection (12). Compounds were classified as either confirmed active (CA, providing protection), confirmed moderately active (CM, not reproducibly providing protection), or confirmed inactive (CI). In our experiments, class CA consisted of 417 compounds, class CM of 1069 compounds, and class CI of 40,282 compounds. The available database (October 1999 Release) contains the screening results for 43,382 compounds.

The aim of this experiment was to find fragments that are, statistically significant, over represented in the active class (CA) and under-represented in the inactive (CI).

The following query was posed to the system: $(freq(f, CA) \geq 13) \wedge (freq(f, CI) \leq 516)$. The thresholds in the queries were determined as follows: The minimum frequency threshold in these queries corresponds to 3 % of the active molecules. In order to determine the maximum allowable frequency in the non-active molecules, we used the X^2-Test applied to a 2×2 contingency table with the class as one variable and the occurrence of the fragment as the other one. In this way, we obtained a maximum frequency of 516 in inactive compounds for the first task, and a maximum frequency of 8 in the moderately actives for the second. Given these frequencies, the occurrence of a fragment in the active class is not due to chance at a significance level of 0.999.

For this task, the total computation time was 19,212.31 CPU seconds (measured in CPU seconds on a Linux PC with a Pentium III 600 MHz processor). The first part (the minimum frequency query) took only 1,544.09 CPU seconds, and the second part (the maximum frequency query) took 17,668.22 CPU seconds. The boundary set *G* contained 222 elements, and *S* contained 314 elements. This contrasts with a total of 1,623 patterns in the solution space bounded by *G* and *S*, which demonstrates the utility of *G* and *S* sets (version spaces) in this kind of application.

In the minimum frequency part of the query, the longest solution fragment had a length of 24 atoms, the longest fragment found in the maximum frequency part had a length of 22 atoms. So, it has been shown that MOLFEA can search for very long patterns in a structural database of over 40,000 compounds.

Table 1.

G	*S*
O-C-n:c:n:c=O	N=N=N-C-C-C-n:c:c:c:n:c=O
O-C-C-C-C-C-n:c=O	N=N=N-C-C-C-n:c=O
C-C-C-O-C-n:c:n:c:c:c	c:c:c:n:c:n-C-C-C-N=N=N
C-c:c:n:c:n:c=O	C-c:c:n:c:n-C-C-C-N=N=N
N-c:c:c-S	C-c:c:n-C-C-C-N=N=N
N-C-C-C-O-C-C-O	C-C-O-C-C-N=N=N
C-C-C-O-C-n:c=O	N=N=N-C-C-O-C-n:c:n:c=O
O-C-n:c:c:c:n:c=O	N=N=N-C-C-O-C-n:c:c:c=O
N=N=N	N=N=N-C-C-C-n:c:n:c=O
N-c:c:c:c:c-s	N=N=N-C-C-C-n:c:c:c=O

From the boundary sets *G* and *S*, we picked fragments based on their class distribution (statistical significance and accuracy). Table 1 summarizes the most significant samples. The majority of these fragments, e.g.

N=N=N-C-C-C-n:c:c:c=O and N=N=N-C-C-C-n:c:n:c=O

indicate compounds that are derivatives of Azidothymidine (AZT, Retrovir, Zidovudine, 3'-Azido-3'-deoxythymidine, CAS 30516-87-1, see Figure 4), a potent inhibitor of HIV-1 replication, which is widely used in the treatment of HIV infection. Other fragments indicate another class of reverse transcriptase inhibitors, mainly thiocarboxanilide derivatives, which are, according to our knowledge, drugs that are still in an experimental phase. The automated rediscovery of the most important classes of anti-HIV drugs indicates the utility of the presented approach

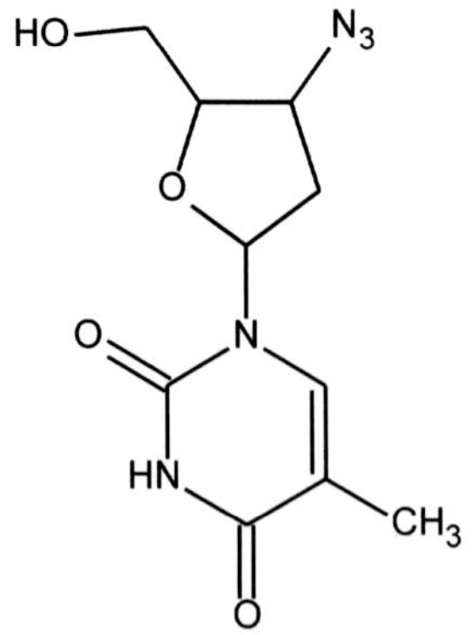

Figure 4. Chemical Structure of Azidothymidine (AZT)

USE OF MOLFEA FEATURES FOR SAR

In a series of other experiments (11), we have employed MOLFEA-generated features in structure-activity relationship prediction. SAR prediction in this context works in three steps. In a first step, MOLFEA queries are used to construct a set of fragments. These queries can be class-blind (e.g. when we require a minimum frequency on the *whole* dataset), or class-sensitive, when we consider separate criteria for active and inactive molecules.

The resulting fragments are used as binary features or fingerprints (a fragment either occurs in a molecule or it does not) to describe the molecules. The resulting data sets can be fed into a traditional data mining system [such as e.g. WEKA (13)], to obtain a predictive model for the biological effect under investigation. This method can be combined with virtually any data mining technique; we have induced decision trees, classification rules, regression models and Support Vector Machines (SVMs).

In the experiments sketched in (11), we have investigated the effects of class-sensitive vs. class-blind fragment construction on benchmark datasets for biodegradability, mutagenicity and carcinogenicity prediction (14, 15, 16). Class-sensitive feature construction is performed using combined minimum and maximum frequency queries as described above. Class-blind feature construction is performed by a simple minimum frequency query.

The predictive accuracies obtained with MOLFEA generated features turned out to be at least competitive with the best published results in the literature so far. Support Vector Machines were able to take advantage of a large number of features (fragments) constructed in a class-blind manner, whereas classical inductive Machine Learning approaches (decision trees and rules) seemed to benefit from class-sensitive feature construction. Summing up, these experiments clearly demonstrated the utility of MOLFEA-generated features in the induction of SARs.

RELATED WORK

Molecular fragments are, among other purposes, useful and important for the the induction of Structure-Activity Relationships. The use of automatically derived structural fragments in SARs originates from the CASE/MultiCASE systems developed by (17).With more than 150 published references, the CASE/MultiCASE systems are the most extensively used SAR and predictive toxicology systems. Previous approaches in these areas are based on the "decomposition" of individual compounds: these methods generate *all* fragments occurring in a given single compound. In this regard, our contribution is a language that enables the formulation of complex queries regarding fragments – users can specify precisely which fragments they are interested in. We also implemented a solver to answer queries in this language. Thus, from the algorithmic point of view, it is no longer necessary to process the results of queries post-hoc.

Molecular fragment finding has also been studied within the context of inductive logic programming and knowledge discovery in databases. For instance, WARMR (18) and the approach by Inokuchi *et al.* (19) have been used in this context. WARMR is a system discovering requently succeeding Datalog queries, and thus is not restricted to fragments. The approach by Inokuchi *et al.* deals with arbitrary frequent subgraphs, and thus is not restricted to linear ragments. Both approaches differ in that their pattern domain is more expressive, but finding requent patterns is likely to be more expensive and complex than for linear fragments.

Finally we want to stress again, thatMOLFEA is only one instance of the general *Inductive Database Framework.* It is quite easy to adapt Inductive Databases for a new application domain. We have presently implemented PROFEA(6), that analyses the secondary structure of proteins and we are working on a further instance for gene expression data.

Table 2.

Domain	Learning Algorithm	Blind	Class Sensitive
Carcinogenicity	C4.5	64.3	65.9
	PART	**67.4**	65.9
	Log	65.6	65.3
	1. SVM	65.3	65.0
Mutagenicity	C4.5	90.4	87.2
	PART	91.5	93.1
	Log	94.7	**95.7**
	1. SVM	94.7	92.0
Biodegradation	C4.5	77.7	76.5
	PART	77.1	79.9
	Log	80.2	75.9
	1. SVM	**81.1**	75.9

FURTHER DEVELOPMENTS

MOLFEA is presently capable to find linear sequences of atoms in databases with chemical structures. It is presently impossible, to identify stereochemical effects, or arrangements in three-dimensional space. We are therefore working on several extensions to the MOLFEA framework.

Abstractions The concept of fragments is not limited to sequences of elements. It is fairly easy to define abstract *atom types*, that can be more general (e.g. H-bond donor/acceptor) or more specific (e.g. oxygen in a carbonyl group) than the elements. Another addition will be the introduction of wildcards for atoms. With these extensions it will be possible to find fragments like "two H-bond donors separated by 5 heavy atoms".

Branched Fragments The extension towards branched fragments is conceptually easy, from a computer science viewpoint, but it might result in increased search times. The use of branched fragments is particularly attractive from a chemist's viewpoint, because with their help it will be possible to identify stereochemical effects.

3D Fragments Another extension of MOLFEA is the consideration of three-dimensional arrangement of atoms in fragment finding. Work on this topic is almost completed and will be the subject of a separate publication.

CONCLUSIONS

We have presented a novel database and data mining approach based on the concept of inductive databases. Even though our framework was presented for string-like patterns, it should be clear that one could easily adapt it towards richer data structures such as e.g. graphs, or towards other application domains in bio- and chemo-informatics. We have also argued that the inductive database framework is useful for knowledge discovery in databases in general and in bio- and chemo-informatics in particular. The authors hope that the work on MOLFEA and PROFEA will stimulate other researchers to add inductive query languages to the many existing databases in bio- and chemo-informatics. This in turn should allow scientists to understand their data more easily and to discover new knowledge more effectively.

ACKNOWLEDGMENTS

This work is partly supported by the European Union IST FET project cInQ.

REFERENCES

[1] Fayyad, U., Piatetsky-Shapiro, G., Smmyth, P. (1996). From Data Mining to Knowledge Discovery: An Overview. In *Advances in Knowledge Discovery and Data Mining*, AAAI Press, Menlo Park, Calif., pp. 1-30.

[2] Imielinski, T. & Mannila, H. (1996). A database perspective on knowledge discovery. *Communications of the ACM* **39**(11):58-64.

[3] Han, J., Lakshmanan, L. V. S., Ng, R. T. (1999). Constraint-based, multidimensional data Mining. *Computer* **32**(8):46-50.

[4] De Raedt, L. (2000). A logical database mining query language. In *Proceedings of the 10th Inductive Logic Programming Conference*, Lecture Notes in Artificial Intelligence, Vol. 1866, Springer Verlag.

[5] Boulicaut, J. F., Klemettinen, M., Mannila, H. (1998). Querying Inductive Databases: A Case Study on the MINE RULE Operator. In *Proceedings of PKDD-98*, Lecture Notes in Computer Science, Vol. 1510, Springer Verlag, pp. 194-202.

[6] Fischer, J. (2002). *Objektorientiertes Design einer induktiven Datenbank und eine Anwendung des Levelwise Version Space Algorithmus auf die Sekundärstruktur von Proteinen*. Studienarbeit at the Machine Learning Lab, University of Freiburg, Germany.

[7] De Raedt, L., Kramer, S. (2001). The level wise version space algorithm and its application to molecular fragment finding. In *Proceedings of the Seventeenth International Joint Conference on Artificial Intelligence*, Morgan Kaufmann.

[8] Kramer, S., De Raedt, L., Helma, C. (2001). Molecular feature mining in HIV data, in: *Proc. of the Seventh ACM SIGKDD International Conference on Knowledge Discovery and Data Mining (KDD-01)*, 136-143.

[9] Agrawal, R., Imielinski, T., Swami, A. (1993). Mining association rules between sets of items in large databases. In *Proceedings of ACM SIGMOD Conference on Management of Data*.

[10] Mannila, H. & Toivonen, H. (1997). Levelwise search and borders of theories in knowledge discovery. *Data Mining and Knowledge Discovery* **1**(3):241-258.

[11] Kramer, S. & De Raedt, L. (2001). Feature construction with version spaces for biochemical applications. *Proceedings of the 18th International Conference on Machine Learning*, 258-265, Morgan Kaufmann.

[12] Weislow, O. S., Kiser, R., Fine, D. L., Bader, J. P., Shoemaker, R. H., Boyd, M. R. (1989). New soluble formazan assay for HIV-1 cytopathic effects: application to high flux screening of synthetic and natural products for AIDS antiviral activity. *Journal of the National Cancer Institute* **81**:577-586.

[13] Witten, I. & Frank, E. (1999). *Data Mining: Practical Machine Learning Tools and Techniques with Java Implementations*. Morgan Kaufmann.

[14] D¡zeroski, S., Blockeel, H., Kompare, B., Kramer, S., Pfahringer, B., Van Laer, W. (1999). Experiments in predicting biodegradability. In *Proceedings of the 9th International Workshop on Inductive Logic Programming (ILP-99)*, 80-91, Springer Verlag.

[15] Srinivasan, A., Muggleton, S., King, R. D., Sternberg, M. (1996). Theories for mutagenicity: a study of first-order and feature based induction. *Artificial Intelligence* **85**(1-2):277-299.

[16] Srinivasan, A., King, R. D., Bristol, D. W. (1999). An assessment of submissions made to the predictive toxicology evaluation challenge. *Proc. of IJCAI-99*, 270-275.

[17] Rosenkranz, H. S., Cunningham, A. R., Zhang, Y. P., Clayhamp, H. G., Macina, O. T., Sussmann, N. B., Grant, S. G., Klopman, G (1999). Development, characterization and application of predictive toxicology models. SAR and QSAR in *Environmental Research*, **10**:277-298.

[18] Dehaspe, L. & Toivonen, H. (1999). Discovery of frequent datalog patterns. In *Data Mining and Knowledge Discovery Journal*, **3**(1):7-36.

[19] Inokuchi, A., Washio, T., Motoda, H. (2000). An Apriori-based algorithm for mining frequent substructures from graph data. In D. Zighed, J. Komorowski, and J. Zyktow (eds.) *Proceedings of PKDD 2000*, Lecture Notes in Artificial Intelligence, Vol. 1910, Springer Verlag.

METABOLIC ANALYSIS IN DRUG DESIGN: COMPLEX, OR JUST COMPLICATED?

ATHEL CORNISH-BOWDEN AND MARÍA LUZ CÁRDENAS

CNRS-BIP, 31 chemin Joseph-Aiguier, B.P. 71,13402 Marseille Cedex 20, France
E-Mail: athel@ibsm.cnrs-mrs.fr

Received: 14th June 2002 / Published: 15th May 2003

ABSTRACT

The metabolism of living organisms is certainly complicated, but it does not follow from this that it is complex, which would mean that its behaviour could not be computed, even in principle. For a simple organism like the parasite *Trypanosoma brucei,* the kinetics of glycolysis can in fact be computed with good accuracy from the known kinetic properties of the component enzymes. There may thus be no complexity to confront in the design of molecules intended to act as drugs. On the other hand, successful drug design will require much more attention to the functions of the intended targets than is evident in current practice, which is overwhelmingly structure-based. This will involve recognizing the different kinds of inhibition possible in a complete system and analysing the stoichiometric constraints that limit the variations in metabolite concentrations that are possible.

INTRODUCTION

Many properties of living systems are complicated, in the sense that the total behaviour can only be accounted for by taking account of many components. Complexity, however, is more than that, implying that the total is in some sense more than the sum of the parts. Rosen (1), for example, defines complexity in the following words: “I have attempted to introduce, and to motivate, a concept of complexity. A system is called complex if it has a nonsimulable model. Above all, complex systems cannot be completely chararacterized in terms of any reductionistic scheme based on simple systems.” Other definitions of complexity exist, for example that of von Neumann, but these are quite similar in meaning and there is a large degree of agreement in modern complexity theory that complexity is something along the lines indicated by Rosen. He goes on to argue that living systems are complex in just this sense, i.e. that a complete description of a living system is, and will always be, beyond the reach of computation. This idea can be traced back to Schrödinger (2), who believed that the study of life would lead to new physical laws, or in other words that the laws of physics as determined from the study of non-living systems would be inadequate to describe life.

Monod (3) and other authors have seen suggestions of vitalism in this idea, so it is important to emphasize that Schrödinger's view is not in any sense vitalism. The essential distinction was made by Rashevsky (4), who pointed that the fact that life is constrained by the laws of physics does not require the existence of life to be predictable from the laws of physics. This parallels the idea that all of chemical kinetics obeys the laws of thermodynamics, but the laws of thermodynamics do not predict kinetic properties; yet a proposition that is regarded as uncontroversial and even trite when comparing thermodynamics with kinetics stimulates accusations of vitalism when it is applied to physics and biology.

Nonetheless, the biological revolution of the past half-century is the fruit of a resolutely reductionist approach, rejecting any idea of "new physics" and assuming that all the properties of living systems can be explained in terms of the physical properties of their components. Moreover, as Savageau (5) pointed out, "any respectable reductionist is also a reconstructionist", by which he meant that once one has characterized all the components of a system one ought to show that the whole system can be reconstructed by putting them all together, adding that "the problem is that the reconstructionist phase is seldom carried out." In this article, therefore, we shall examine a metabolic system, that of glycolysis in *Trypanosoma brucei,* where not only have the individual enzymes been well characterized kinetically, but the properties of the entire system can indeed be calculated from the properties of the components. If this is typical then it suggests that even though drug design may be complicated, requiring many different things to be taken into account, it is not complex. However, even if this is agreed, it does not justify using the same word to mean something else. Using "complex" to mean complicated is ultimately as confusing as it would be if biochemists agreed that vitalism in the 19th century sense had no part to play in biochemistry, but then decided to use the word "vitalism" to mean enzyme catalysis.

A separate problem derives from the fact that although lip-service is paid to structure-function relationships in attempts to rationalize drug discovery, the effort in practice is overwhelmingly devoted to structure, i.e. to the search for molecules with structures complementary to the known structures of biological molecules. Function is, if not completely forgotten, then at least given no emphasis, and metabolism, an essential aspect of biological function, may pass unmentioned in entire issues of journals devoted to drug discovery (6), or mentioned only rarely, and then only in the context of the metabolic transformation of drugs, ignoring the metabolic functions of the drug targets (7). Although biotechnology is often presented as if progress in the past two decades represented a major success, the reality is different. For example, ten major classes of antibiotics were discovered between 1935 and 1963, but after 1963 there has been just one, the oxazolidones. This sort of observation explains the pessimism of some recent commentators (8, 9) on the state of the drug industry.

To illustrate the possibilities of doing better by taking account of the real behaviour of metabolic systems, we can examine how one might modify the activity of an enzyme in the cell (for example by genetic manipulation, or by the action of an inhibitor, etc.) to satisfy a technological aim. For example, if the objective is to eliminate a pest, one might suppose that the effect of an inhibitor could be to depress an essential flux to a level insufficient for life, or to raise the concentration of an intermediate to a toxic level. The former may seem the more obvious, but the latter is easier to achieve in practice, and there are some excellent examples of industrial products that work in that way, such as the herbicide glyphosate and antimalarials of the quinine class. A study of glycolysis in the parasite *Trypanosoma brucei* (which causes African sleeping sickness) indicates that for this approach to work the selected target enzyme must have a substrate with a concentration that is not limited by stoicheiometric constraints. That is not necessarily easy to find in a complicated system, and typically needs the metabolic network to be analysed in the computer.

IMPORTANCE OF INHIBITION TYPE

A high proportion of drug targets are enzymes, and in consequence a high proportion of drugs are enzyme inhibitors. The question therefore arises of how different types of enzyme inhibition affect the potential for pharmacological effects. From the kinetic point of view linear inhibitors range from competitive to uncompetitive inhibitors, separated by a large class of mixed inhibitors (including pure non-competitive inhibitors, the special case with identical competitive and uncompetitive components). This classification is complicated by the fact that irreversible inhibitors are sometimes confused with pure non-competitive inhibitors, because when added in doses insufficient to inactivate the enzyme completely they decrease the concentration of active enzyme while leaving the properties of the still-active molecules unchanged. Tight-binding inhibitors are also easy to confuse with pure non-competitive inhibitors because whatever the true inhibition type the inevitable slowness of the inhibitor-release step means that in experiments on a short time scale they are indistinguishable from irreversible inhibitors. This classification is illustrated in Fig. 1.

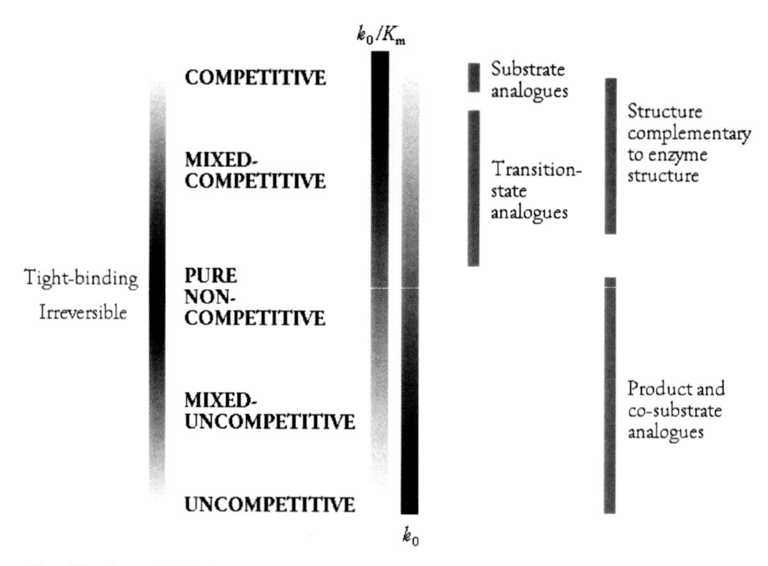

Figure 1. Classification of inhibitors. From the kinetic point of view simple inhibitors range from competitive, with effects only on the apparent value of the specificity constant k_0/K_m, to uncompetitive, with effects only on the catalytic constant k_0. There is a large class of mixed inhibitors with effects on both, including pure non-competitive inhibition, in which the competitive and uncompetitive components are equal. Tight-binding inhibitors may in principle fall into any of these types, but in practice are often made difficult to distinguish from pure non-competitive inhibitors by the very slow release of inhibitor in what is theoretically a reversible binding. Similar considerations apply to true irreversible inhibitors. The Figure also illustrates the types of inhibitory behaviour expected for various kinds of inhibitor that can be designed on the basis of structural considerations.

The inhibition types of various molecules likely to arise from purely structural considerations are also shown in the figure, but before discussing these we need to examine how different kinds of inhibition affect the kinetic behaviour of a living cell.

The first and possibly most important point is that there may be no easily observable effect at all unless the enzyme activity is decreased by a large factor, and perhaps not even then if the metabolic function of the inhibited enzyme can be replaced by an alternative enzyme or series of reactions. The typical form of the dependence of metabolic flux on the activity of any enzyme is as illustrated in Fig. 2, and in practice most enzymes are located in normal conditions near or to the right of the point labelled B, not near the point labelled A; in other words variation of an enzyme activity around its normal value will typically have little or no effect on the metabolic flux. This is not merely a theoretical expectation (10) but it is also confirmed by numerous experimental studies (11, 12).

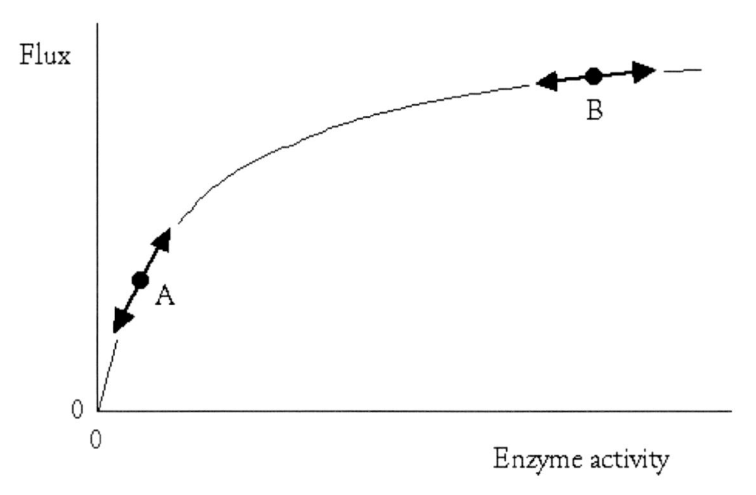

Figure 2. Dependence of metabolic flux on enzyme activity. The curve shown is typical of those found in numerous experimental cases (11, 12), and in the normal state nearly all enzymes are located near or to the right of the point labelled B; extremely few are found near the point labelled A. This implies that varying an enzyme activity *in vivo* will typically have no perceptible effect on the metabolic flux.

Many gene knock-out studies have produced no observed phenotypic effects: in *Saccharomyces cerevisiae,* for example, more than 80% of genes are "silent" (13), in the sense that any of them can be suppressed with no effect on growth or other gross aspects of the phenotype. Although this sort of result apparently surprised many observers it was entirely predictable and expected, as it follows almost automatically from results from metabolic control analysis that have now been in the literature (10) for nearly 30 years. The essential point is that a typical enzyme has a flux control coefficient close to zero for a gross phenotypic property like growth, and even for a more specific property such as the flux through the pathway in which the enzyme is located the flux control coefficients of most enzymes are small. This means that even if an enzyme is known to play an essential role in processes relevant to a particular disease, there is no certainty that inhibiting it will have a significant effect on the disease unless the inhibition is very strong (For activators the situation is even worse: as flux control coefficients normally decrease when the activity of an enzyme is increased, it is rare almost to the point of non-existence for activation of an enzyme *in vivo* to have a perceptible effect on the flux through the enzyme.).

The common inhibition types are easily confused in experiments in the spectrophotometer, with the result that cases of mixed inhibition are frequently reported as competitive. However, ordinary steady-state experiments, typically done at substrate and product concentrations decided and fixed by the experimenter, are very misleading as a model of inhibition *in vivo,* where concentrations are not fixed at all, and certainly not by an external agent such as an experimenter. For a typical enzyme that catalyses a reaction in the middle of a metabolic pathway it is a better approximation (though still not exact) to consider that the rate is fixed and that the substrate and products are adjusted by the enzymes that use them to whatever values will sustain the appropriate flux. In these conditions competitive and uncompetitive inhibition become very different from one another (14), and the uncompetitive component becomes the main determinant of the response of the system to a mixed inhibitor.

These points are illustrated in Fig. 3 for the system of ten enzymes shown in Fig. 3a. When the inhibition is competitive (Fig. 3b) effects on both flux and metabolite concentrations are very slight, but all become much larger when the inhibition is uncompetitive (Fig. 3c).The essential point is that a molecule that competes with a substrate is a molecule that a substrate can compete with, and so the effect of a competitive inhibitor can be nullified by relatively minor adjustments of the concentrations around the inhibited enzyme. By contrast, effects of uncompetitive inhibitor are potentiated by the variations in substrate that they generate, and fairly modest levels of inhibition may therefore produce huge changes in substrate concentration. It is essentially this kind of effect that is exploited by glyphosate, an inhibitor of 3-phosphoshikimate 1-carboxyvinyltransferase, uncompetitive with respect to 3-phosphoshikimate (15).

Although the results of Fig. 3 are quite general for the effects of inhibiting an enzyme in the middle of a pathway, not all enzymes are located in the middles of pathways, and circumstances exist in which the illustration may be misleading. If an enzyme catalyses the first step in the transformation of a substrate such as glucose that is maintained at a stable and relatively high concentration by regulatory mechanisms independent of the pathway of interest, then it will resemble an enzyme in a spectrophotometer with a fixed substrate concentration rather than the sort of enzyme considered in Fig. 3. Another complication can arise when certain metabolite concentrations are constrained to remain within definite limits by stoichiometric relationships. Nonetheless, there remain many enzymes for which Fig. 3 gives a realistic picture of the likely effects of inhibition.

We now return to the part of Fig. 1 that is concerned with enzyme inhibitors designed on the basis of structural considerations, i.e. substrate analogues, transition-state analogues, molecules

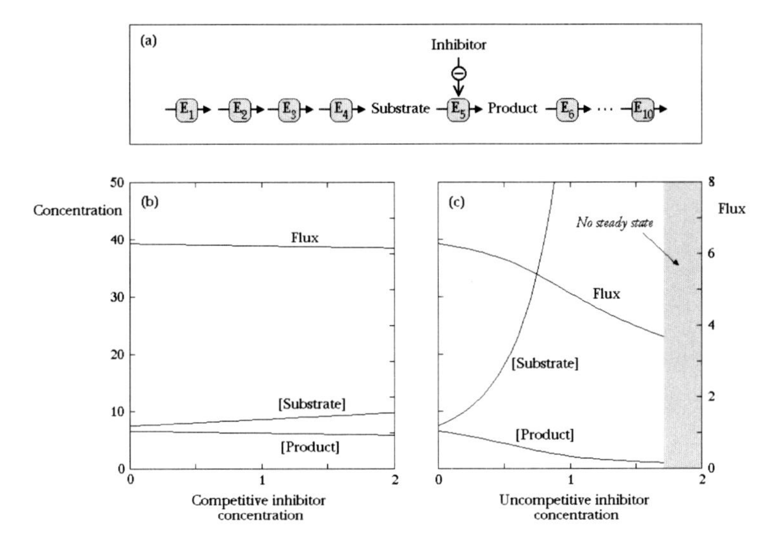

Figure 3. Effects of inhibition on an enzyme in the middle of a long pathway. (**A**) The pathway includes an enzyme E_5 that is the fifth of a series of ten enzymes and is acted on by an external inhibitor. Details of the assumptions used are given elsewhere (21); qualitatively they do not affect the results. (**B**) If the inhibition is competitive the effects both on the flux through the pathway and on the concentrations of the substrate and product of the inhibited enzyme are slight if the concentration of inhibitor does not become very large compared with the inhibition constant. (**C**) However, if the inhibition is uncompetitive the effects, especially on the substrate concentration become much larger, and can result in complete loss of the steady state. In (**B**) and (**C**) the inhibitor concentrations are given relative to the inhibition constants.

with structures complementary to parts of the enzyme, and analogues of products or co-substrates. Substrate analogues provide an easy choice: it is sufficient to find a molecule that resembles the substrate sufficiently to have similar binding properties, but which lacks a property essential for reaction to occur, as in the classic example of malonate, a structural analogue of succinate that acts as a competitive inhibitor of succinate dehydrogenase. Unfortunately, however, such a molecule is almost certain to act as a simple competitive inhibitor and is unlikely to bind appreciably more tightly than the substrate. As Fig. 3b shows, therefore, the easiest solution is unlikely to be the best.

A transition-state analogue is normally also predominantly competitive, though it may have significant uncompetitive character if the substrate of interest is the second or later substrate to bind in a sequential process. Moreover, it also often binds much more tightly than the substrates, and so it may be possible to deliver it to the binding site at a concentration much higher than its inhibition constant, and a significant effect can then be expected even for a competitive inhibitor. Molecules designed to bind to specific structural features of a target enzyme will in general have similar characteristics to transition-state analogues, with the same advantages and disadvantages. However, careful choice of a molecule that can bind when the substrate is bound but not otherwise may in principle result in an uncompetitive inhibitor, which may be expected to have a large pharmacological effect, but we are not aware of any examples where this has been done in practice: the need for the inhibitor to bind only when the substrate is bound is commonly ignored in attempts to design drugs.

The final case to be considered is the one exemplified by the herbicide glyphosate (often sold as "Roundup"), whose chemical name is *N*-phosphonomethylglycine. Comparison of its structure with those of 3-phosphoshikimate and phospho*enol*pyruvate (Fig. 4), the two substrates of 3-phosphoshikimate 1-carboxyvinyltransferase, makes it clear that it is not a structural analogue of the first of these, though it could more plausibly be regarded as an analogue of the second. Kinetically, it does not bind to the free enzyme but it does compete with phospho*enol*pyruvate for the enzyme-3-phosphoshikimate complex. The authors who reported this finding found it surprising, and noted that corresponding competition is *not* found between *N*-phosphonome-thylglycine and phospho*enol*pyruvate in other apparently analogous cases, such as pyruvate kinase (15).

The example thus illustrates several points relevant to the search for uncompetitive inhibitors. A molecule that does not resemble the metabolite considered to be the substrate for the target enzyme but which does resemble one of its co-substrates may well prove to be uncompetitive with respect to the substrate of interest. (Of course, in the context of enzyme mechanisms there is no difference between a substrate and a co-substrate. Each is as much a substrate as the other. However, in the metabolic context the distinction is both commonplace and, usually, meaningful.).

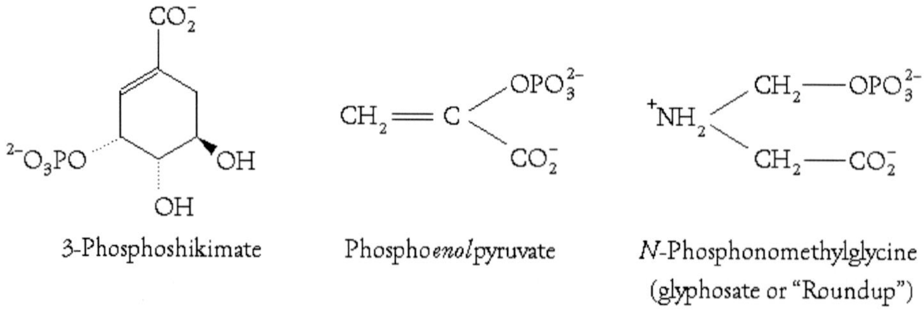

3-Phosphoshikimate | Phospho*enol*pyruvate | *N*-Phosphonomethylglycine (glyphosate or "Roundup")

Figure 4. Comparison of the structure of *N*-phosphonomethylglycine (glyphosate, or "Roundup") with those of the two substrates of 3-phosphoshikimate 1-carboxyvinyltransferase.

Unfortunately, however, the co-substrate of any particular target enzyme will often be a metabolite like NAD, ATP or, as in this case, phospho*enol*pyruvate, that is also a substrate for other enzymes. Designing an uncompetitive inhibitor as a close structural analogue of a co-substrate then incurs the risk that it will lack specificity: even if it has the desired effect on the target enzyme it will also have similar but undesired effects on other enzymes. *N*-Phosphonomethylglycine then appears as a very fortunate case: similar enough to phospho*enol*pyruvate to interact strongly with one enzyme, but not similar enough to interact with others.

Designing an inhibitor with significant and specific uncompetitive character is thus a much more difficult task than designing a competitive inhibitor, because it cannot just be a substrate analogue. This difficulty is not an adequate reason for not attempting it, however, because solving a difficult task is likely to be more rewarding than solving an easy task if its solution is potentially useful and the solution to the easy problem potentially useless.

SYSTEMIC CONSIDERATIONS

It should never be forgotten that apart from a few secreted enzymes like invertase, enzymes do not act *in vivo* in isolation but as components of systems. This is the crucial point that explains why experiments in the spectrophotometer are often a poor guide to what is likely to happen *in vivo*. It is essential for considering fluxes, because the low flux control coefficients of most enzymes mean that it is normally very difficult to decrease a flux significantly by inhibition, and almost impossible to increase one by activation or overexpression (Fig. 2). Even when the objective is not to vary a flux but to vary a metabolite concentration the systemic context of the inhibition remains relevant because there are at least two circumstances where uncompetitive inhibition may not be much more effective than competitive. One of them we have already mentioned: an enzyme that acts on glucose at the beginning of a minor pathway, for example, will have very little effect on the glucose concentration in any ordinary conditions, because that is determined by controls on the major glucose-using pathways like glycolysis and glycogen synthesis; such an enzyme can therefore be treated like an enzyme in a spectrophotometer, and will respond to competitive inhibition as readily as to uncompetitive inhibition.

The other point is that a metabolite concentration can only show a large response to changes in the activities of enzymes that consume it or produce it if it is largely free from stoichiometric constraints. Some constraints are obvious from inspection: for example, in a cell with a fixed total NAD concentration the concentrations of neither reduced nor oxidized NAD can exceed the fixed total. However, much more complicated constraints may also exist, and identifying these may require stoichiometric analysis by computer.

Simulating the Metabolism of *Trypanosoma Brucei*

The bloodstream form of the parasite *Trypanosoma brucei* has an unusually simple metabolism, and most of its metabolic activity is shown in Fig. 5. As most of the enzymes and transporters involved have been purified and their kinetic behaviour thoroughly characterized, it was possible to set up a detail kinetic model of the system as illustrated, and this was done by Bakker and colleagues (16). Both they and later we (17) found that the computer model was able to reproduce the known behaviour of the living organism with a fair degree of accuracy, and our suggestion on the basis of the model that inhibiting export of pyruvate would be lethal to the parasite proved (unknown to us at the time) to correspond to reality (18). All of this suggests that trypanosomal metabolism may be complicated (at least compared with ordinary kinetic experiments in the spectrophotometer even if it is far from complicated by comparison with other organisms), but that it is not complex.

As the pyruvate transporter is probably not the most obvious choice of target for drug design, given the nearly 20 other apparently reasonable targets in the metabolic scheme, we shall briefly indicate how it followed from the computer analysis. It is obvious from inspection of Fig. 5 that there are three simple conservation constraints, representing the sum of the two forms of NAD in the glycosome, the sum of the three adenine nucleotides in the glycosome, and the sum of the three adenine nucleotides in the cytosol. These last two are separate because these species do not cross the glycosomal membrane. In addition to these three constraints there is a fourth, which was identified by computer analysis; this involves all the phospho-groups in the glycosome that are not accounted for by entry of inorganic phosphate or exit of 3-phosphoglycerate, as well as two additional phosphorylated intermediates that are partitioned between the glycosome and the cytosol (19). In Fig. 5 these phospho-groups are marked in such a way as to make the conservation as obvious as possible, but if the structures are shown in a more conventional way the conservation relationship is far from obvious.

It follows from this analysis that in *Trypanosoma brucei* nearly all of the metabolites in the whole system participate in conservation relationships.

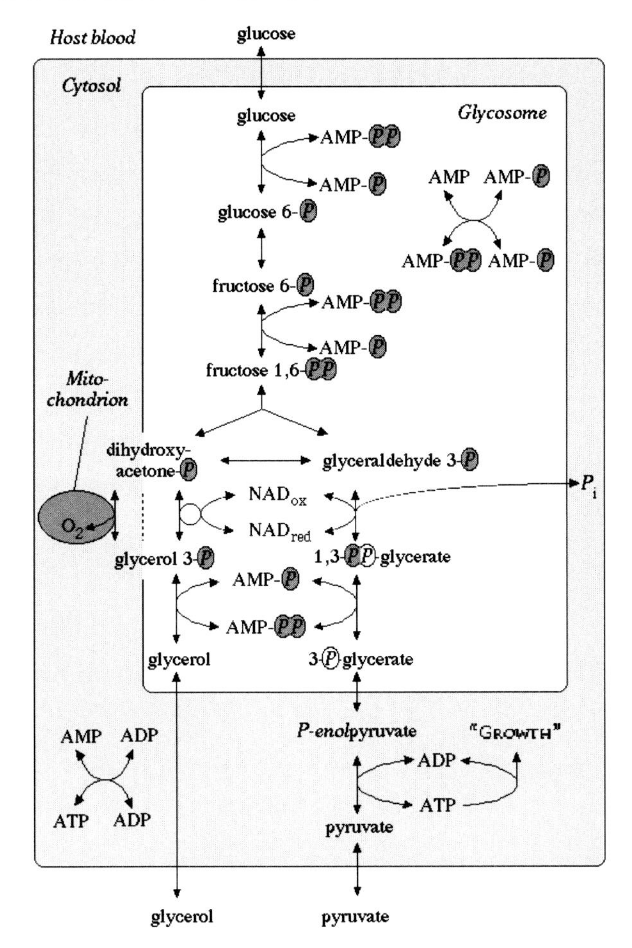

Figure 5. The glycolytic pathway in bloodstream form *Trypanosoma brucei*. There are four compartments, labelled Host blood, Cytosol, Glycosome and Mitochondrion. Dihydroxyacetone phosphate and glycerol 3-phosphate diffuse between the glycosome and the cytosol, but the two transport steps are not explicitly shown; glycerol 3-phosphate is reoxidized under aerobic conditions to dihydroxyacetone phosphate on the membrane of the mitochondrion. "GROWTH" represents all of the steps in the rest of metabolism that are driven by dephosphorylation of ATP. The phospho groups involved in the complicated conservation relationship are shown as *P* against a shaded background; the two unconserved phospho groups in the glycosome are shown against a white background.

This means that their concentrations cannot be changed by large amounts by altering enzyme activities, and this in turn means that extremely few of the enzymes and transporters in the system are plausible targets for a drug intended to act like glyphosate in plants. One of these few is the pyruvate transporter, and it appeared to be the only one that could not be eliminated for other reasons (17). Thus despite the apparent multitude of plausible drug targets in Fig. 5, there is only one that survives a closer analysis of the system.

DISCUSSION

The model of *Trypanosoma brucei* that we have considered is far from being a complete model of a living organism, as it ignores many processes, such as protein synthesis, cell division etc. that are certainly essential for the life of the organism even if they account for a relatively minor part of the metabolic activity of the parasite. Unfortunately this is no less true of all other metabolic models currently being studied, most of which include less experimental information and cover a much smaller proportion of the total metabolism of the cell than the *Trypanosoma brucei* model does. The question therefore remains open as to whether setting up a model that will allow computer simulation of an entire organism is possible, even theoretically, and thus whether living systems are truly complex in the sense used in complexity theory. However, both our results (17) and those of others (16) indicate that a large part of the trypanosomal metabolism that is likely to be of interest for drug design can certainly be simulated, with computer-generated results that correspond closely with experimental observations. In this limited sense, therefore, trypanosomal metabolism is not complex, and we expect that this will prove to be true for many other cases of pharmacological interest, and thus there will be no complexity to confront. On the other hand there are certainly complications that need to be taken into account, and progress in drug design will continue to be slow until the existence of enzymes *in vivo* as components of systems and not as entities on their own is generally recognized. This was understood by Wright (20) as long ago as 1934, but remains poorly understood by the world at large even today.

REFERENCES

[1] Rosen, R. (2000). *Essays on Life Itself,* 306, Columbia University Press, New York.

[2] Schrödinger, E. (1992). *What is Life?* pp. 76–85, Cambridge University Press, Cambridge (originally published 1944).

[3] Monod, J. (1970). *Chance and Necessity,* pp. 27–28, Collins, London.

[4] Rashevsky, N. (1972). *Organismic Sets,* pp. 18–19, Mathematical Biology Inc., Holland, MI.

[5] Savageau, M. A. In *Control of Metabolic Processes* (Cornish-Bowden, A.; Cárdenas, M. L., eds.), pp. 69–87, Plenum, New York

[6] *Nature* (1996), **384** (suppl.):1.

[7] *Science* (2000), **288**:1951.

[8] Horrobin, D. F. (2001). *Nature Biotechnol.* **19**:1099.

[9] Cárdenas, M. L. & Cornish-Bowden, A. (2000). *Science* **288**:618.

[10] Kacser, H. & Burns, J. A. (1973). *Symp. Soc. Exp. Biol.* **27**:65.

[11] Flint, H. J., Tateson, R. W., Barthelmess, I. B., Porteous, D. J., Donachie, W. D., Kacser, H. *Biochem. J.* **200**:231

[12] Fell, D. (1997). *Understanding the Control of Metabolism,* pp. 136–150, Portland Press, London.

[13] Cornish-Bowden, A. & Cárdenas, M. L. (2001). *Nature* **409**:571.

[14] Cornish-Bowden, A. (1986). *FEBS Lett.* **203**:3.

[15] Boocock, M. R. & Coggins, J. R. (1983). *FEBS Lett.* **154**:127.

[16] Bakker, B. M., Michels, P. A. M., Opperdoes, F. R., Westerhoff, H. V. (1997). *J. Biol. Chem.* **272**:3207.

[17] Eisenthal, R. & Cornish-Bowden, A. (1998). *J. Biol. Chem.* **273**:5500.

[18] Wiemer, E. A. C., Michels, P. A. M., Opperdoes, F. R. (1995). *Biochem. J.* **312**:479.

[19] Cornish-Bowden, A. & Hofmeyr, J.-H. S. (2002). *J. Theor. Biol.,* in press.

[20] Wright, S. (1934). *Amer. Nat.* **63**:24.

[21] Cornish-Bowden, A. & Cárdenas, M. L. (2002). *Compt. rend. Acad. Sci.,* in press.

Molecular Informatics: Confronting Complexity, May 13th - 16th 2002, Bozen, Italy

aMAZE: A Database of Molecular Function, Interactions and Biochemical Processes

Christian Lemer[1], Avi Naim[2], Yong Zhang[2], Didier Croes[1], Georges N. Cohen[4], Gaurab Mukherjee[2], Lorenz Wernisch[2,3], Klaudia Walter[2], Jean Richelle[1], Jacques van Helden[1] and Shoshana J. Wodak [1,2]*

[1]Centre de Biologie Structurale et Bioinformatique, CP160-16, Université Libre de Bruxelles, 50 Av. F. Roosevelt, 1050 Bruxelles, Belgium

[2] European Bioinformatics Institute (EBI), Genome Campus – Hinxton, Cambridge CB10 1SD, UK

[3] School of Crystallography. Birkbeck College. University of London. Malet Street. London WC1E 7HX, UK

[4] Unité d' Expression des Gènes Eucaryotes,Institut Pasteur, 28, rue du Docteur Roux, 75524 Paris Cedex 15, France

E-Mail:* Shosh@ucmb.ulb.ac.be

Received:15th July 2002 / Published: 15th May 2003

Abstract

The aMAZE database (http://www.amaze.ulb.ac.be) manages information on the molecular functions of genes and proteins, their interactions and the biochemical processes in which they participate. Its data model embodies general rules for associating molecules and interactions into large complex networks that can be analysed using graph theory methods. The processes represented include metabolic pathways, protein-protein interactions, gene regulation, transport and signal transduction. These processes are mapped into their spatial localisation. A distinct feature of aMAZE is its Object-Oriented, modular and open user interface. Queries are invoked through dedicated modules, data can be linked to external sources, interactively browsed and transferred between modules, and new modules can be readily added. Available modules also include, a custom-built Diagram Editor for the automatic layout, display, and interactive modification of pathway diagrams, and procedures for analysing network graphs.

INTRODUCTION

A major challenge of the post genomic era is to determine the biological function of all the genes and gene products in the growing number of newly sequenced genomes, and to understand how they interact to yield a living cell. To meet this challenge, experimental efforts of unprecedented magnitude are being undertaken for investigating gene expression patterns, analysing the entire protein complement of cells and characterising the full repertoire of protein-protein interactions. These efforts involve high-throughput techniques, such as micro-array based gene expression analysis (1,2), two-hybrid screens (3-6), and large scale protein characterisation (7). All these techniques generate very large amounts of data that must be interpreted in an iterative bootstrapping approach in light of the available information on the molecular and cellular function of genes.

This has brought to the forefront, the pressing need for more efficient systems for managing and analysing complex information on biological function. Public databases such as SWISS-PROT (8), are a rich source of information on function, but they store it mostly in text form, not readily amenable to computer analysis.

Efforts have therefore been undertaken to develop more specialised databases, for representing information on cellular processes and interactions (see reference 9 for a recent review). Some databases deal primarily with metabolic pathways (10-12). Others focus on protein-protein interactions (13,14), on gene regulations (15,16) or on signal transduction (16,17).

The aMAZE database system presented here, implements a data model, which embodies general rules for associating individual biological entities and interactions into large complex networks of cellular processes (9,18). It can deal with a large variety of cellular processes comprising metabolic pathways, protein-protein interactions, gene regulation, sub-cellular localisation, transport, and signal transduction. The major aim of this system is to provide a general open framework for: 1) combining information from different levels of cellular organisation, 2) flexibly querying and visualising this information, 3) custom-building of tools for advanced programmatic analyses, and 4) greatly facilitating annotation of data on complex cellular processes. The aMAZE system or others like it, should help the biologists in understanding, analysing, and ultimately modelling, complex cellular networks.

THE AMAZE DATA MODEL

In aMAZE data are organised using an Object Oriented model, as described previously (9,18). This model distinguishes between two fundamental classes of objects, *BiochemicalEntity* and *Interaction*. The first represents physical entities (protein, gene, compound, etc.), with attributes pertaining to structural properties (polypeptide sequence, gene position on the chromosome). The second represents molecular activities, which can be of several types. *EntityProcessing* and *Binding,* which are interactions having entities as input and as output (e.g. chemical reaction, protein-protein interaction), and *Control*, which are interactions having an entity as input and another interaction as output (e.g. a catalysis in a relationship between a protein and a reaction).

A third important class in aMAZE is *Process*, which represents a collection of interconnected process elements. These elements consist either of individual interactions or of entire processes. Using this representation, graphs of biochemical pathways can be reconstructed by linking the interactions through their inputs and outputs. Higher level views, for example, of the interconnections between different biochemical pathways (pathways of pathways), can also be generated.

As discussed previously (18), our model has the great advantage of defining the activities of a particular structural entity (compound, gene or protein) within a given context, rather than within the entity object itself, thereby allowing for a flexible description of multiple activities of individual genes and proteins.

The description of localisation, a central issue in representing biological function, is also handled in aMAZE. This is done using the class *Compartment*, which is further subdivided into the sub-classes, *SubcellularCompartment, CellType, Tissue, Organ,* and *Systematic group.* Any *Process* can be linked to a given combination of objects in the *Compartment* class in order to describe where it occurs (e.g. plasma membrane of T-cells in *Homo sapiens*).

The aMAZE data model also comprises various types of classification schemes. One type is the containment hierarchies, e.g. the nucleosome is contained in the nucleus, itself contained in the cell. Another type concerns various classifications of objects of the aMAZE model. Those include the systematic classification of organisms, sub-cellular compartments, compounds, and so on. The ability of representing independently and simultaneously one or more of the so-called biological Ontologies (19), which contain functional classifications, is also provided. Finally, the aMAZE database also includes its own meta description - classes used to represent the aMAZE data model itself - which can hence be queried and analysed like any other information stored in the database.

IMPLEMENTATION ASPECTS

The aMAZE system is implemented in a mixed Object-Oriented/Relational environment. The data are manipulated as objects, but stored in Tables using a relational DBMS (Oracle). The Object-Relational mapping is performed using an algorithm that converts the full object description into the relational schema, in ways that can be readily understood by humans, while preserving key properties of the object model, such as inheritance and polymorphism (Lemer et al., to be published). This conversion also enables the user to choose between the Object-Oriented or relational modes, according to need, without loss of user friendliness or query power.

The server architecture comprises three layers: the Service, Access and Client layers. The Service layer manages multiple access to the relational database for query purposes and updates. The Access layer manages the users and the access rights, and is also in charge of load balancing between servers. The Client layer manages network communication and provides the API (Application Programmatic Interface) to the server layer. The different layers are developed in Java and are connected via the Remote Method Invocation system (RMI).

THE aMAZE FRAMEWORK: A MODULAR OBJECT-ORIENTED OPEN USER INTERFACE

User access to aMAZE is provided via the aMAZE_Framework. This is a multi-document application, where each type of query to the database is invoked through a dedicated module, as illustrated in figure 1, and data can be transferred between modules using *drag & drop* operations.

Query modules currently available comprise the retrieval of molecular entities (gene, protein small molecule) by name, and by requiring an exact or partial match. They also allow the retrieval of different types of interactions, each defined by one or more inputs, one of more outputs, or a list of both inputs and outputs. Another module allows the selection of objects on the basis of any feature (attribute) of interest. This module can be used iteratively to refine a selection by sequential filtering.

Upon query completion, the user can drag and drop the results into the Object-Holder, which enables viewing the results list. Individual items of the list can moreover be dropped into the Object-Viewer, in order to visualise their complete description (see Fig. 1). Importantly, retrieved entities or interactions can in turn be entered as input into a new query window via the drag and drop facility.

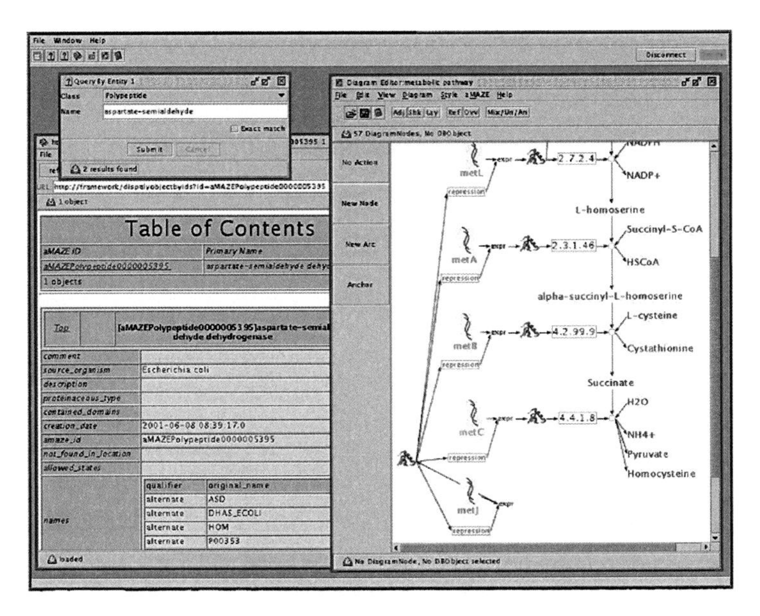

Figure 1. Screen shot of the aMAZE-Framework in action. The upper left corner shows the query by entity module, where a polypeptide (aspartate-semialdehyde dehydrogenase) is retrieved using only part of its name (aspartate-semialdehyde) as input. The information on the retrieved polypeptide object is displayed using the Object-Viewer, which appears as the large application window visible directly beneath the query window. The Diagram Editor, window, displaying a pathway appears of the right half of the Figure. It shows part of the methionine biosynthesis pathway. The different functions available in the Diagram Editor appear can be activated by the icons displayed on the tool bars on the top and the left hand sides. Catalysis interactions are displayed by boxes containing the corresponding EC number. Small molecule compounds are displayed in purple, proteins appear as blue icons, all of which show the polypeptide backbone of a small protein. These can be replaced by picture of the individual proteins, once the link with the Protein Databank (PDB) is established. The 'exp' and 'repression' objects represent the expression and repression interactions, respectively. Gene names are in green.

SQL queries to the relational database can also be submitted directly by invoking a specialised module. Users unfamiliar with SQL, can run pre-canned custom queries, prepared in advance with help from the database administrator, and accessed via the SQL module. The results of the SQL query can in turn be dropped into any other display module (Object-Holder, Object-Viewer, Diagram-Editor). Thus, the framework combines the power of SQL queries with the user-friendliness of an object representation.

At any given time, multiple windows can remain open, facilitating data traffic between them. The Object-Viewer can furthermore be used as a general HTML WEB browser, thereby enabling direct links to external databases. The latter aspect is extremely useful for all operations that may benefit from analysing information from different sources. The aMAZE-Framework has been developed in 'pure' Java, and has already been tested on different platforms (Solaris, Linux, Windows, Macintosh).

THE DIAGRAM EDITOR

A key module of the Framework is the Diagram Editor. This is a custom built graphical editor, which allows to display diagrams of cellular processes retrieved from the database (e.g. pathway diagrams), and to interactively modify them to suit particular aesthetic preferences. Modified diagrams can be saved locally in different formats (jpg, png, text ..), printed, and (under some conditions) re-submitted to the database.

In addition to a number of useful standard display options, users can chose to collapse specific nodes or node types, with or without their complete set of anchored descendents. Modified diagrams can be stored locally, to be printed or displayed subsequently.

Options for an automatic layout of pathway diagrams are also provided. This is performed using a custom-built multi-directional hierarchical layout algorithm, tailored for the representation of biochemical pathways (van Helden, unpublished), as illustrated in figure 1.

When the Diagram Editor is integrated into the Framework, any set of displayed objects can be dragged and dropped into a query or Object-Viewer document. We expect this to be particularly helpful, as it enables the biologist to manipulate complex data in an intuitive fashion. It should likewise be very instrumental for entering new information on complex biological processes, as the newly entered information can not only be displayed graphically but also validated for consistency against information already stored in the database. In addition if a newly entered process involves entities or interactions already stored in aMAZE, work can be saved by using the stored information.

GRAPH ANALYSIS TOOLS

Another group of modules features a set of tools for analysing cellular processes algorithms adapted from graph theory (20). They include for example, finding the shortest path or N-shortest paths, between a given pair of source and target nodes. In performing these operations a limit on path length (number of intervening nodes) can be imposed to limit the number of generated solutions, and thus reduce calculation time.

Path enumeration is not restricted to annotated pathways but can be performed on the global metabolic network built on the fly from all the chemical reactions and compounds stored in the database, irrespective of where (organism, compartment) the corresponding catalytic reactions were observed. This has many useful applications. A typical application is, given a cluster of co-expressed genes (identified in DNA micro-array experiments), find the shortest paths linking together, in a biologically meaningful way, most or all of the activities carried out by the cluster members (9,18). This can help in assigning gene function, in identifying alternatives to classical pathways and in discovering new pathways. Figure 2a, b illustrates the method, and Figure 3 its application to the reconstruction of the path of chemical reactions linking the activities carried out by the genes from the so-called Met cluster in the experiments of Spellman et al. (1998) (19). A detailed description of the path-finding approach and some more general issues in representing networks of metabolic processes can be found in reference 20.

Another application is to characterise the functional interactions between pairs of enzyme coding genes found to be involved in fusion/fission analyses (22-25). To that end, the path enumeration algorithms are used to compute the length of the shortest path between to corresponding pair of catalyzed reactions. The shorter the path, the greater the likelihood that the two enzymes functionally interact (20).

The performance of theses algorithms is currently being tested by applying them to rebuild known metabolic pathways, starting from different subsets of catalysed reactions (20). Results suggest that the algorithms should also be very useful for the on-going work on pathway annotation, as pathways for which all reactions and compounds are already stored in the database can be rebuilt and displayed automatically before examination by the annotator.

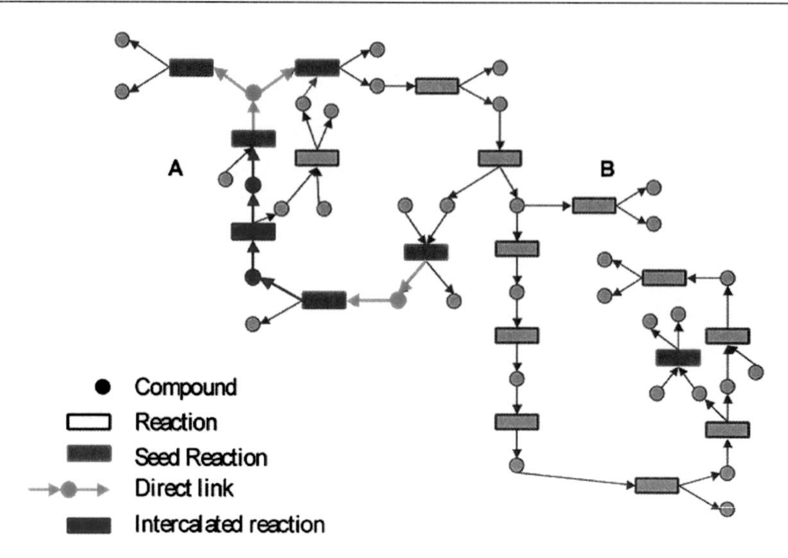

Figure 2. Reconstruction of a metabolic pathway from a set of seed reactions.

A) This figure schematises a putative network of chemical reactions linked via their substrates and products (filled circles), which are small molecule compounds being transformed into one another by the chemical reactions (filled rectangles). Information on all the reactions of the network is stored in the database. Red rectangles indicate chemical reactions given as seeds to the graph analysis program. The program then attempts to links these seed reactions via their substrates and products. When a direct link (light greed arrows and circles) cannot be made, the program is allowed to intercalate 1 or more reactions, which were not part of the initial set of seeds (dark green arrows, circles and rectangles).

B) Sub-graph extracted using the procedure described in **A)**.This sub-graph represents the pathway built from the set of initial seed reactions. Note that one seed reaction (right hand side of Figure) could not be connected, and is therefore discarded from the analysis. Comparing the constructed pathway with known pathways stored in the database can then be then used to determine if the pathway is already known, is a variant of a known pathway, or a completely new one.

DATA CONTENT, AND ANNOTATION EFFORTS

The main sources of the data in aMAZE are, the BRENDA database (27), containing information on the majority of classified enzyme reactions and the proteins that catalyse them, the KEGG database (12), and SWISS-PROT (8). For data on pathways, we completely rely on in-house annotations.

Presently, aMAZE contains information on more than 150,000 genes, about 200,000 polypeptides, over 8000 chemical reactions, a similar number of small molecule compounds, and about 60 pathways of metabolic regulation in *E.coli* and *S. cerevisiae*. Information on over 150 additional pathways has been collected and will be entered into the aMAZE system by the fall of 2002. These will include all the known metabolic regulation and signal transduction pathways of *E.coli* and most such pathways in *S. cerevisiae*.

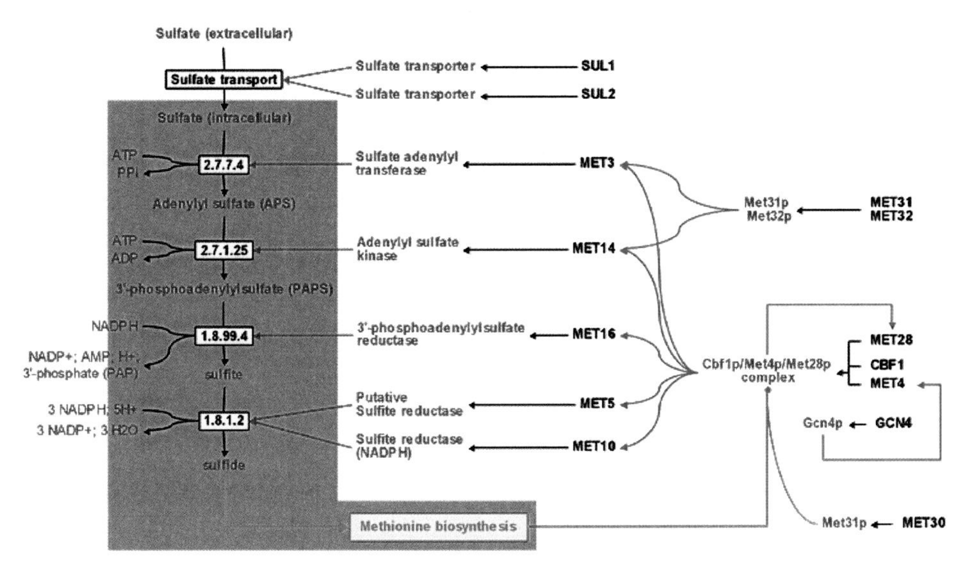

Figure 3. Pathway reconstruction from the cluster of cell-cycle regulated genes (19), involved in sulfur assimilation in *S. Cerevisiae*.

A) Part of the reconstructed pathway that involves the enzyme coding genes marked in black (Met3, Met5, Met10, Met14, Met16). This part deals with sulfate to sulfite transformation.

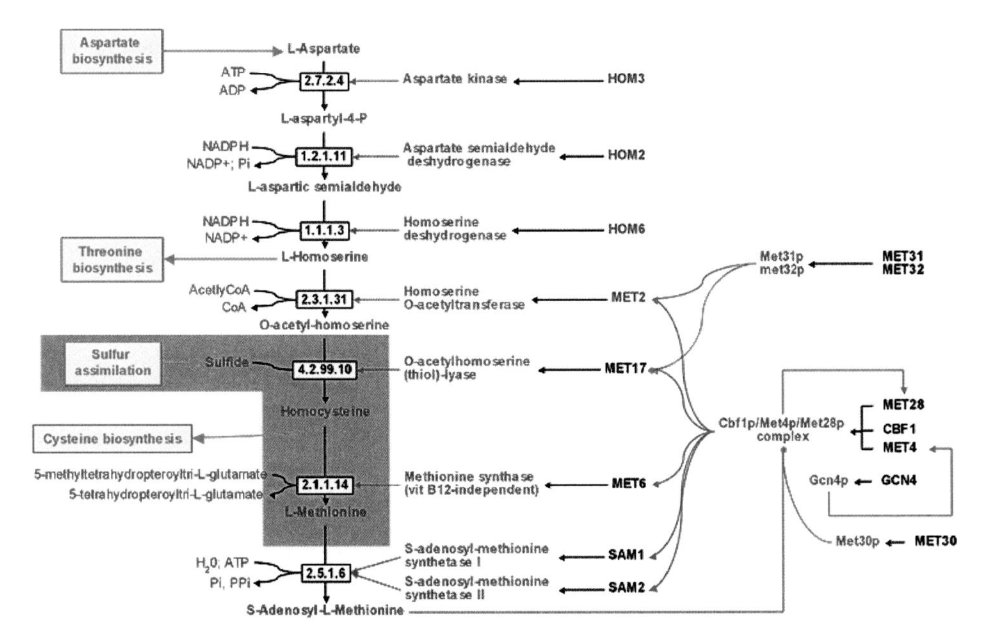

B) Part of the reconstructed pathway, which belongs to the classical pathways of methionine biosynthesis. It involved the enzyme coding genes marked in black (Met6, Met17). The method used to reconstruct the pathway is the one illustrated in Fig. 2. The genes used as seeds for the pathway reconstruction belong to those from the so-called Met cluster in Spellman's experiments (19).

The flexible query and graphical tools available in aMAZE, together with the specialised annotation modules that are currently being developed (see below), should greatly facilitate the annotation efforts in the very near future. Groups interested in annotating pathways will be provided with access to aMAZE, as soon as these modules become functional, and we very much hope that such groups will contact us. As with all information in aMAZE, contributed data are processed and assigned unique references, enabling the identification of the annotator for the record, as well as for query purposes.

FUTURE DIRECTIONS

A major priority is to further populate the database with available information on cellular processes and data on protein-protein interactions in the yeast *S. cerevisiae*, which is among the best characterised eukaryotic systems so far. This will allow evaluating the full potential of aMAZE on a consistent dataset, and should form the basis for extending the annotation efforts to higher eukaryotes, or less well characterised prokaryotes. In the near future, this effort will focus on processes pertaining to cell division and cancer, in mouse and human.

A second priority is the expansion of the query capabilities. In particular we will focus on the application of graph analysis tools to the interpretation of gene expression and protein-protein interaction data on terms of cellular pathways, and tools for visual and quantitative comparison of pathways and networks.

Much work also remains to be done on the representation of data on the small molecule compounds, so that queries can be formulated on selecting and comparing compounds and reaction on the basis of the compound chemical structure and sub-structure. This information should also allow the introduction of criteria based on chemistry for the analysis and construction of metabolic pathways.

AVAILABILITY AND ACCESS

The aMAZE system will be freely accessible to academic research groups over the Web starting the fall of 2002. The available functionalities will include all query modules, and protocols for custom development of new query methods. All the data in aMAZE, except for those belonging to third parties or declared as confidential, are freely available. It is also our intention to make the entire aMAZE system publicly available starting April 2003, date at which the commitments towards the consortium of Industries supporting this project is ending. Groups interested in developing their own applications are most welcome to contact us. We would also be happy to provide specialised groups, or individuals, interested in contributing annotations on cellular pathways, with privileged access to the custom annotation modules of aMAZE, as soon as those are available.

ACKNOWLEDGEMENTS

We thank Thure Etzold, and David Gilbert for useful discussions. Many of the ideas for the aMAZE data model are due to Renato Mancuso, and the first prototype version of the database was implemented by Matthew Eldridge. Kirill Degtyarenko is thanked for valuable help with data annotation. This work was sponsored in part by a consortium of industries, comprising Astra-Zeneca, Aventis, Monsanto, Organon, and Roche. We thank scientists from these companies for valuable input. Support from the European Commission (contracts QLRI-CT-1999-01333 and BIO4-CT98-0226), is also gratefully acknowledged.

REFERENCES

[1] Brown P. O. & Botstein, D. (1999). Exploring the new world of the genome with DNA microarrays. *Nat. Genet.* **21**:33-37.

[2] DeRisi J. L., Iyer, V. R., Brown, P. O. (1997). Exploring the metabolic and genetic control of gene expression on a genomic scale. *Science* **278**:680-686.

[3] Fields, S. & Song, O. (1989). A novel genetic system to detect protein-protein interactions. *Nature* **340** (6230):245-6.

[4] Rain, J. C., Selig, L., De Reuse, H., Battaglia, V., Reverdy, C., Simon, S., Lenzen, G., Petel, F., Wojcik, J., Schachter, V., Chemama, Y., Labigne, A., Legrain, P. (2001). The protein-protein interaction map of Helicobacter.

[5] Uetz, P., Giot, L., Cagney, G., Mansfield, T. A., Judson, R. S., Knight, J. R., Lockshon, D., Narayan, V., Srinivasan, M., Pochart, P., Qureshi-Emili, A., Li, Y., Godwin, B., Conover, D., Kalbfleisch, T., Vijayadamodar, G., Yang, M., Johnston, M., Fields, S., Rothberg, J. M. (2000). A comprehensive analysis of protein-protein interactions in *Saccharomyces cerevisiae*. *Nature* **403** (6770):623-7.

[6] Ito, T., Chiba, T., Ozawa, R., Yoshida, M., Hattori, M., Sakaki, Y. (2001). A comprehensive two-hybrid analysis to explore the yeast protein interactome. *Proc. Natl. Acad. Sci. U S A* **98** (8):4569-74.

[7] Fields, S. (2001). Proteomics. Proteomics in genomeland. *Science* **291** (5507):1221-4.

[8] Bairoch A. & Apweiler, R. (2000). The SWISS-PROT protein sequence database and its supplement TrEMBL in 2000. *Nucleic Acids Res.* **28**:45-48.

[9] van Helden, J., Naim, A., Mancuso, R., Eldridge, M., Wernisch, L., Gilbert, D., Wodak, S. J. (2000). Representing and analysing molecular and cellular function using the computer. *J. biol. Chem.* **381** (9-10): 921-35.

[10] Karp, P. D., Riley, M., Saier, M., Paulsen, I. T., Paley, S. M., Pellegrini-Toole, A. (2000). The EcoCyc and MetaCyc databases. *Nucleic Acids Res.* **28**:56-59.

[11] Overbeek, R., Larsen, N., Pusch, G. D., D'Souza, M. Jr. E. S., Kyrpides, N., Fonstein, M., Maltsev, N., Selkov, E. (2000). WIT: integrated system for high-throughput genome sequence analysis and metabolic reconstruction. *Nucleic Acids Res.* **28**:123-125.

[12] Kanehisa, M. & Goto, S. (2000). KEGG: Kyoto Encyclopedia of Genes and Genomes. *Nucleic Acids Res.* **28**:27-30.

[13] Xenarios, I., Rice, D. W., Salwinski, L., Baron, M. K., Marcotte, E. M., Eisenberg, D. (2000). DIP: the database of interacting proteins. *Nucleic Acids Res.* **28** (1):289-91.

[14] Bader, G. D., Donaldson, I., Wolting, C., Ouellette, B. F., Pawson, T., Hogue, C. W. (2001). BIND : The Biomolecular Interaction Network Database. *Nucleic Acids Res.* **29** (1):242-5.

[15] Huerta A. M., Salgado, H., Thieffry, D., Collado-Vides, J. (1998). RegulonDB: a database on transcriptional regulation in *Escherichia coli. Nucleic Acids Res.* **26**:55-59.

[16] Takai-Igarashi, T., Nadaoka, Y., Kaminuma, T. (1998). A database for cell signaling networks. *J. Comput. Biol.* **5**:747-754.

[17] Wingender, E., Chen, X., Hehl, R., Karas, H., Liebich, I., Matys, V., Meinhardt, T., Pr, M., Reuter, I., Schacherer, F. (2000). TRANSFAC: an integrated system for gene expression regulation. *Nucleic Acids Res.* **28**:316-319.

[18] van Helden, J., Naim, A., Lemer, C., Mancuso, R., Eldridge, M., Wodak, S. (2001). From molecular activities and processes to biological function. *Briefings in Bioinformatics* **2** (1):98-93.

[19] Spellman, P. T., Sherlock, G., Zhang, M. Q., Iyer, V. R., Anders, K., Eisen, M. B., Brown, P. O., Botstein, D., Futcher, B. (1998). Comprehensive identification of cell cycle-regulated genes of the yeast *Saccharomyces cerevisiae* by microarray hybridization. *Mol. Biol. Cell* **9** (12):3273-97.

[20] van Helden, J., Wernisch, L., Gilbert, D., Wodak, S. J. (2002), Graph-based analysis of metabolic networks. Ernst Schering Research Foundation Workshop 39, Bioinformatics and Genome Analysis.

[21] Mewes, H. W., Seidel, H., Weiss, B. Editors, *Springer* 245-274.

[22] Rison, S. C. G., Hodgman, T. C., Thornton, J. M. (2000). Comparison of functional annotation schemes for genomes. *Funct. Integr. Genomics* **1**:56-69.

[23] Marcotte, E. M., Pellegrini, M., Ng, H. L., Rice, D. W., Yeates, T. O., Eisenberg, D. (1999a). Detecting protein function and protein-protein interactions from genome sequences. *Science* **285** (5428):751-3.

[24] Marcotte, E. M., Pellegrini, M., Thompson, M. J., Yeates, T. O., Eisenberg, D. (1999b). A combined algorithm for genome-wide prediction of protein function. *Nature* **402** (6757):83-6.

[25] Enright, A. J., Iliopoulos, I., Kyrpides, N. C., Ouzounis, C. A. (1999). Protein interaction maps for complete genomes based on gene fusion events. *Nature*. **402** (6757):86-90.

[26] Tsoka, S. & Ouzounis, C. A. (2000). Prediction of protein interactions: metabolic enzymes are frequently involved in gene fusion. *Nat. Genet*. **26** (2):141-2.

[27] Schomburg, D., Salzmann, D., Stephan, D. (1990-1995). *Enzyme handbook*. 13 vols, Springer.

Hits, Leads, & Artifacts from Virtual & High Throughput Screening

Brian K. Shoichet, Susan L. McGovern, Binqing Q. Wei, and John J. Irwin

Northwestern University, Department of Molecular Pharmacology & Biological Chemistry, 303 E. Chicago Ave, Chicago, IL 60611-3008, USA

E-Mail: b-shoichet@northwestern.edu

Received: 31st July 2002 / Published: 15th May 2003

Molecular docking attempts to find complementary fits for two molecules, typically a candidate ligand and a macromolecular receptor. Among the most popular applications of docking computer programs is that of screening a database of small molecules for those that might act as ligands for a biological receptor of known or modeled structure. The motivating idea is that the receptor structure can act as a template to select database molecules that will complement it structurally and chemically, and so bind to it, modulating its function. The hope is that this will allow novel families of ligands to be found, allowing one to escape from the tedium of substrate-based or analog-based design (Figure 1).

Although simple in principle, docking screens are shot through with uncertainty. Even small molecule ligands have several rotatable bonds, six is not uncommon, and the receptor site has many more. The number of conformations to be explored in docking rises exponentially with the rotatable bonds, so that even for a small molecule ligand this can be a daunting problem. Whereas most docking programs sample small molecule flexibility, the protein is often left rigid. There are some reasons, moreover, to worry that introducing conformational flexibility into the enzyme could, if not done carefully, make docking performance worse, not better (1).

If sampling is challenging, ranking the database molecules for fit in the site is harder still. Calculating absolute binding energies for a protein and a small molecule ligand is notoriously difficult even for very detailed, time consuming techniques, such as Free Energy Perturbation (FEP).

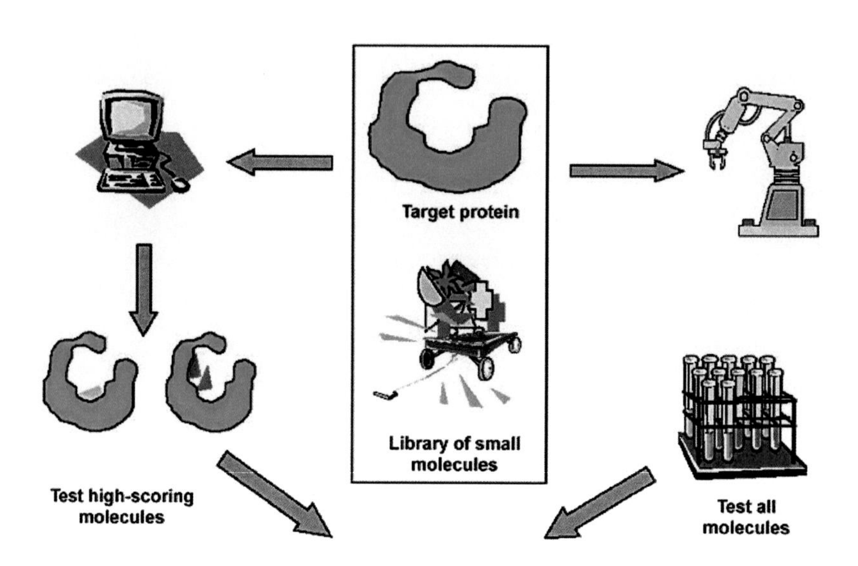

Figure 1. Docking (left) and high throughput screening (right) to discover new leads for drug discovery.

In docking a database of 10^5 to 10^6 small molecules, one cannot afford the time devoted to FEP nor can one afford the assumption that one will be able to compare similar molecules—the databases are purposefully diverse, often maddeningly so. Thus we must make breathtaking assumptions to calculate docking energies or, as they are often (and more honestly) called, docking scores. Our force-fields are inaccurate, the role of solvent is difficult to model (2), we do not relax our systems and therefore do irreversible work, charges are poorly modeled and don't polarize, and we massively under-sample. Getting absolute binding energies from docking calculations is currently well beyond the field. Even monotonic rankings are untrustworthy. Database docking is best considered a screening process, that in favorable circumstances can enrich possible true ligands and filter out unlikely ligands. Like experimental screens, docking screens are plagued by false positives and false negatives.

An appropriate question is why go through the bother of docking at all? Why not just use high throughput methods to *experimentally* screen a database of molecules? Surely this would avoid all the ambiguities of docking and discover more compounds to boot?

Here we consider three related projects ongoing in our laboratories at Northwestern University that consider several of these problems. To investigate how well docking might do at predicting new compounds and their geometries, we first consider a very simple binding site, one that avoids many of the problems that one usually faces in docking. This cavity site in T4 lysozyme is in some senses a "perfect" docking site, since it is so simple. We then consider how well docking does when compared to a HTS project against the same target. These were studies performed in collaboration with Doman and colleagues at Pharmacia, and consider hit rates and quality of hits using both docking and HTS against the enzyme Protein Tyrosine Phosphatase 1B, a diabetes target (3). Finally, we turn to consider a class of promiscuous inhibitors that appear as "hits" from both virtual and high throughput screens. Through a series of biophysical experiments we seek to define a common mechanism of action for a broad range of small molecule non-specific "inhibitors" that have turned up over the years from screens. These nuisance compounds are among the biggest practical problems in using screening for drug discovery research.

A CAVITY BINDING SITE IN T4 LYSOZYME.

In 1991, Matthews and colleagues introduced a cavity into the hydrophobic core of T4 lysozyme by the substitution Leu99→Ala (L99A) (4). This left a completely hydrophobic cavity of about 150 Å^2 in size. As it happened, this site was able to bind small, typically aryl, hydrocarbons in sizes that ranged from benzene, towards the lower end, to naphthalene towards the upper end (Figure 2). Through the work of Morton and Baase (5, 6), over 50 ligands were found that bound to this site, and nine of them were characterized crystallographically.

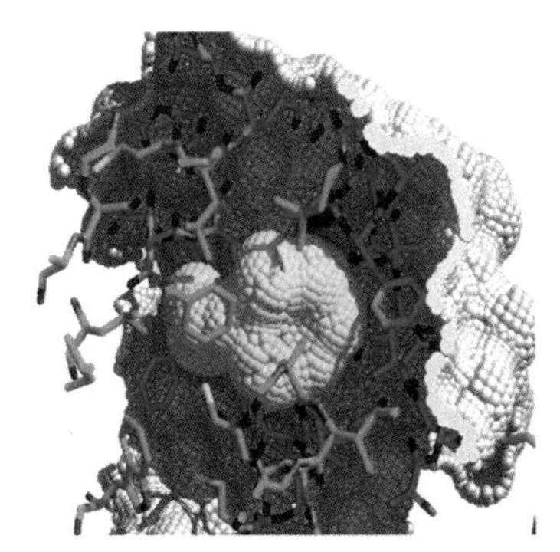

Figure 2. Two views of cavity site in the mutant T4 lysozyme L99A. Outer protein surface in gray, inner cavity surface in yellow. The right panel shows a cutaway of the site, revealing benzene bound in its crystallographic orientation.

We first asked how well docking the Available Chemicals Directory (ACD), which contained most of the characterized ligands for this site, would do at predicting known ligands, using the Northwestern University version of DOCK [Kuntz, 1982 #35; Ewing, 1997 #1107] (NWU

DOCK) (7, 9). As we moved from simple, steric-based scoring to more sophisticated energy and solvation-corrected methods, molecular docking was better and better at enriching known ligands from among the ~170,000 decoys in the database (Figure 3). The best enrichment came when we moved to calculating partial atomic charges and solvation energies for the database molecules using semi-empirical quantum-mechanics through the program AMSOL (10).

Having found that we could retrospectively reproduce known ligands for L99A, we turned to prospective prediction.

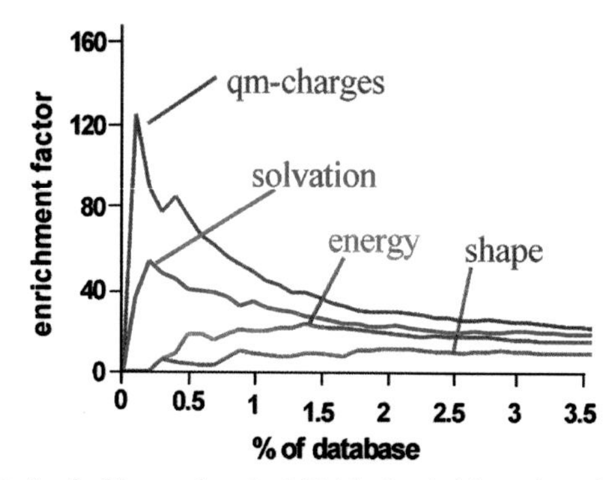

Figure 3. Enrichment plots for docking against the L99A hydrophobic cavity using different scoring functions (Wei et al., submitted for publication).

We substituted one of the hydrophobic residues that line the cavity, Met102, with a more polar glutamine (L99A/M102Q). X-ray crystallography suggested that this substitution introduced a single polar atom, the Oe1 of now Gln102, into the cavity surface. We re-docked the ACD against this slightly polar site, and looked for molecules that: a. scored better against L99A/ M102Q than they did against L99A; b. ranked better in the L99A/M102Q screen than they did in L99A screen; and c. were not observed to bind to L99A site experimentally. Seven molecules were picked and tested for binding; all seven were observed to bind to L99A/M102Q. Five of these were tested in detail using isothermal titration calorimetry (ITC), and were found to have dissociation constants in the 100 μM range (Table 1).

To investigate how well the predicted docked structure of these new compounds corresponded to experiment, the structure of the complexes of five of these compounds was determined by x-ray crystallography, to between 2.0 and 1.85 Å resolution. Before structure determination, predictions were sent to our collaborators in the Matthews lab (Larry Weaver & Walt Baase) to make it a fair test. For all structures, the docking predictions corresponded to the experimental result to with 0.4 Å rms (Figure 4).

Table 1. Binding data for L99A/M102Q		
ligand	**ΔTm (K)[a]**	**K_d (μM)[b]**
3-methylpyrrole	2.1	160
3-chlorophenol	2.7	56
2-fluoroaniline	1.8	100
2,4-difluoroaniline	1.69	
phenole	2.25	91
2,4-difluorophenol	1.9	
3,5-difluoraniline	1.75	
toluene	0.5	160

[a] Binding measured from Tm upshift.
[b] ITC binding data.

In this simple site, molecular docking can predict novel ligands and do so with high geometric accuracy. Perhaps more importantly, the cavity sites L99A, L99A/M102Q, and other derivatives, provide good model systems for testing future developments in docking programs. Docking has advanced to a point where there is a need for model systems that allow both retrospective and prospective testing.

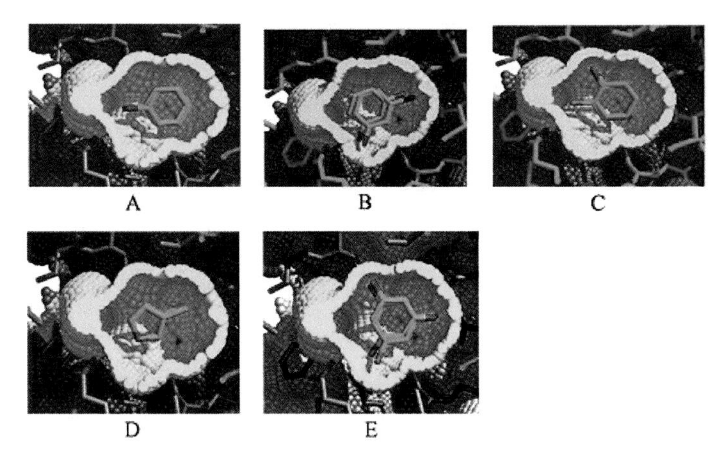

Figure 4. Correspondence between docked (carbons in cyan) and crystallographic configurations of novel ligands in the L99A/M102Q binding site. A: Phenol, B: 3-chlorophenol, C: 2-fluoroaniline, D: 3-methylpyrrole, E: 3,5-difluoroaniline.

DOCKING VS. HIGH THROUGHPUT SCREENING

It's one thing to find that docking can make predictions in what amounts to a "toy" site, but how does it do against a real drug target, and how does it compare to the dominant tool used in the pharmaceutical industry for discovery research, high throughput screening?

This question cannot be answered definitively by any single project, on which caveats will always hang like scabby mendicants. In the spirit of comparing virtual to high throughput screening in as head-to-head manner as possible, we were pleased to collaborate with Doman and colleagues at Pharmacia in their effort to discover novel inhibitors of the Type II Diabetes target PTP1B. At Pharmacia, an in-house library of about 400,000 compounds was screened by HTS.

At Northwestern, about 250,000 commercially available molecules (most from the ACD) were screened using NWU DOCK against the structure of PTP1B (11). About 1000 high scoring compounds were selected by our group at Northwestern, and of these the Pharmacia group chose 365 to actually purchase and test. The results from these 365 compounds were compared to the results from the 400,000 compounds tested experimentally by HTS. All compounds were tested at Pharmacia by Pharmacia biochemists.

The hit rate resulting from docking was 1,700-fold better than the hit rate from HTS (Table 2) (3). More absolute inhibitors were found by testing 365 dock-derived molecules than were found from testing 400,000 compounds from HTS. Surprisingly, the dock-derived inhibitors were more drug-like than the HTS hits (Figure 5). Intriguingly, there was no overlap between the docking and the HTS hits, even at the chemical similarity level, when the two groups of hits were clustered. This last observation suggests that virtual and high throughput screening are complementary techniques; the high hit rate enhancement from docking, should it turn out to be general, suggests that virtual screening is not uncompetitive with HTS.

The thoughtful reader might ask themselves why so many HTS hits were non-drug like? There are several answers to this question, but among them is that many screening hits are artifactual. This is a horrible problem for early drug discovery, because these nuisance compounds can overwhelm true ligands that might exist in one's hit lists. The mechanistic bases of one class of these artifacts is the subject of our last section.

Table 2. Hit rates from docking and high throughput screening against PTP1B.

Technique	Compounds tested	Hits with $IC_{50} < 100\ \mu M$	Hits with $IC_{50} < 10\ \mu M$	Hit Rate
HTS	400,000	85	6	0.021%
Docking	365	127	18	34.8%

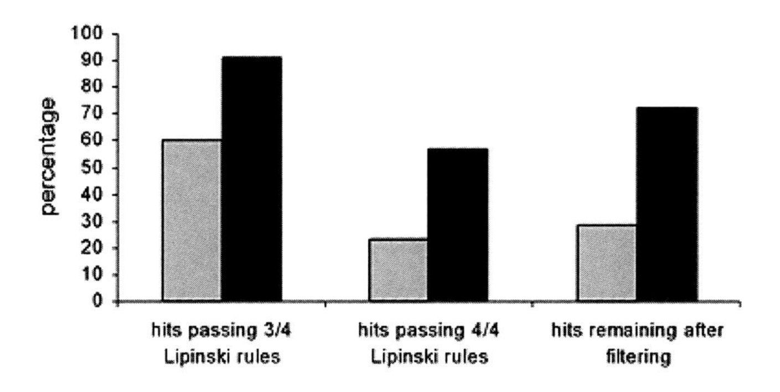

Figure 5. Drug like qualities of PTP1B HTS (diagonal lines) and docking (solid bars) hits, inhibiting at the 100 μM level. Filtering was performed at Pharmacia using internal rules (3).

PROMISCUOUS INHIBITORS FOR VIRTUAL AND HIGH THROUGHPUT SCREENING

We backed our way into this problem, not meaning to. We had undertaken a docking screen against AmpC b-lactamase, an enzyme with which we have a great deal of experience as an experimental system-enzymology, stability, and crystallography are all well in hand. We found tens of novel micromolar inhibitors for this enzyme, which was at first gratifying. To test specificity, we did counter screens against other enzymes including chymotrypsin, trypsin, dihydrofolate reductase (DHFR), malate dehydrogenase (MDH) and b-galactosidase. All of the b-lactamase inhibitors we had discovered turned out to be inhibitors, to varying degrees, of these other, unrelated enzymes (Table 3) (12).

We wondered how widespread this phenomenon of promiscuous inhibition was. We looked through the literature for virtual or HTS hits that looked, vaguely, like the ones we had seen for AmpC. Those that were commercially available we tested against our panel of model, out-group enzymes: AmpC, chymotrypsin, DHFR (or MDH) and b-galactosidase. Many of these compounds inhibited these model enzymes (Table 3).

The inhibition properties were unusual. All of these molecules showed time dependent, but apparently reversible inhibition. Inhibition was very sensitive to ionic strength. Wondering if these compounds were acting as denaturants, we looked to see if urea or guanidinium improved inhibition. Just the opposite happened, inhibition got worse. Similarly, inhibition was very sensitive to the presence of albumin (BSA), which at the 1mg/ml level dramatically attenuated inhibition.

Table 3. Nonspecific inhibitors discovered by screening (12).

Structure	Original Target(s)	IC_{50} (μM) β-lactamase	Chymotrypsin	cDHFR	β-gal
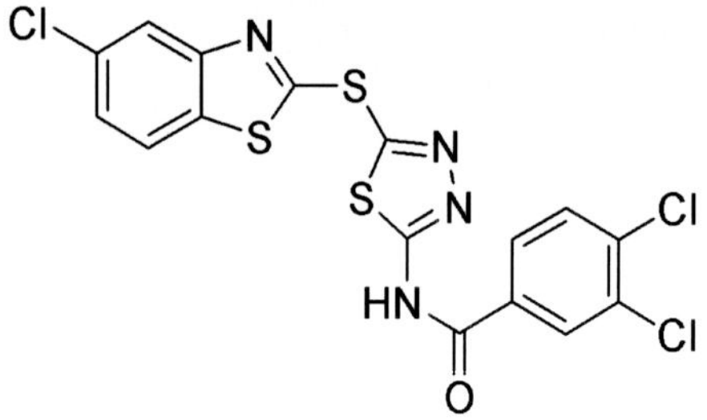	0.5 β-lactamase[a]	0.5	2.5	5	15
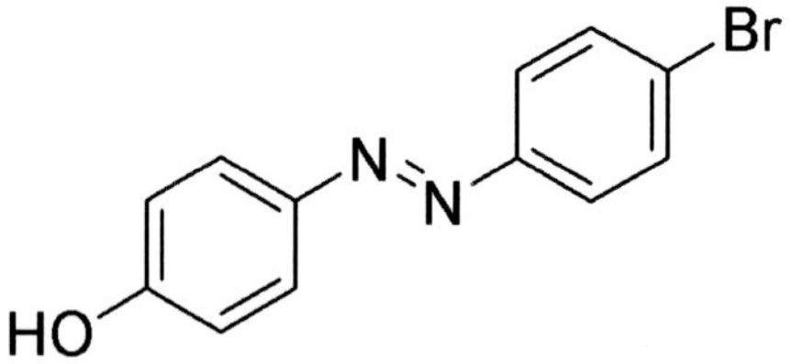	5 β-lactamase[a]	5	25	35	90
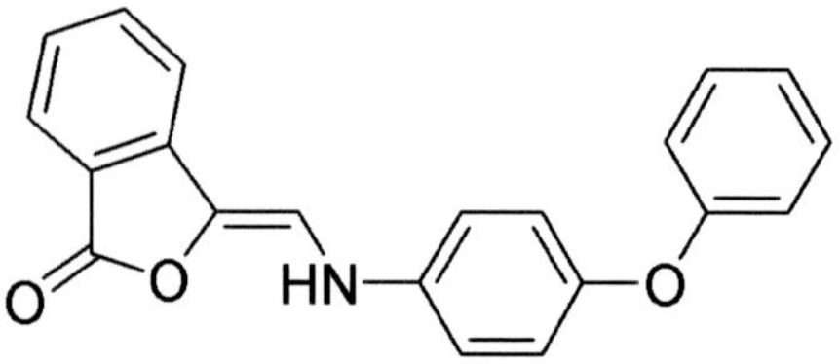	5 β-lactamase[a]	5	15	N.D.	N.D.
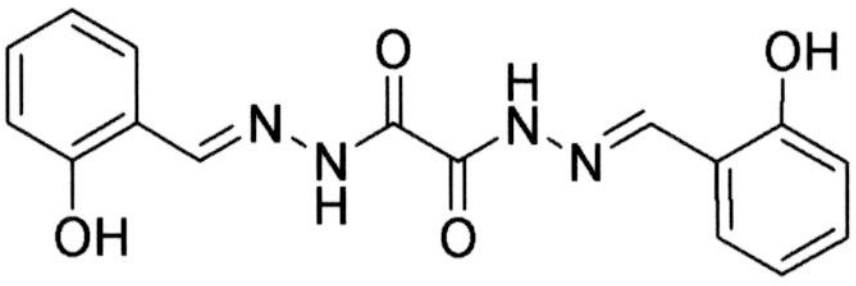	8 malarial protease	10	55	70	180
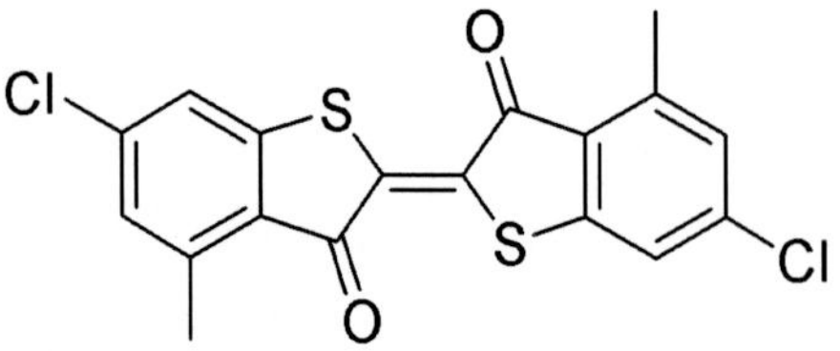	7 pDHFR	10	50	60	300

Table 3, continued. Nonspecific inhibitors discovered by screening (12).

Structure	Original Target(s)	IC_{50} (μM) β-lactamase	Chymotrypsin	cDHFR	β-gal
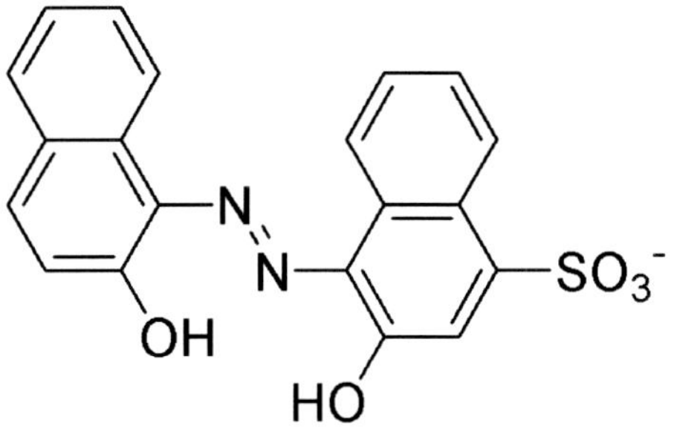	80 pDHFR	50	25	N.D.	600
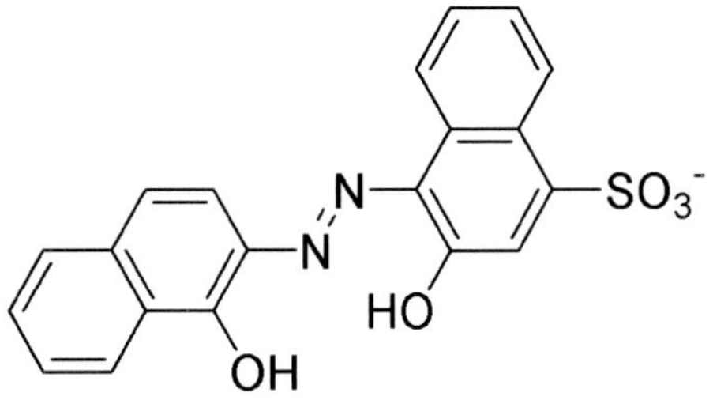	50 HIV Tar RNA	10	90	N.D.	600
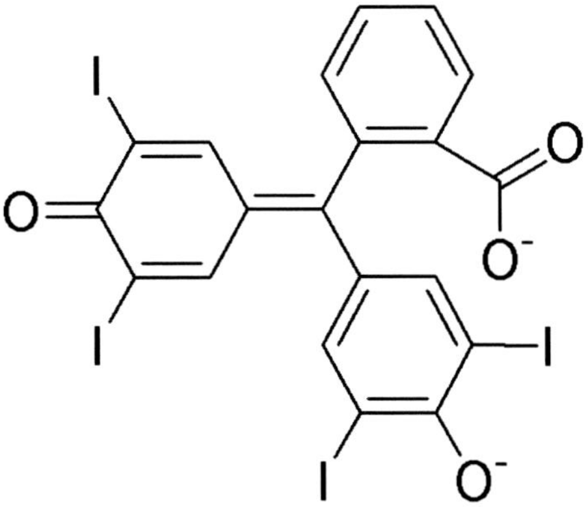	3 TS, 30 kinesin	3	11	20	200

Table 3, continued. Nonspecific inhibitors discovered by screening (12).

Structure	Original Target(s)		IC$_{50}$ (μM) β-lactamase	Chymotrypsin	cDHFR	β-gal
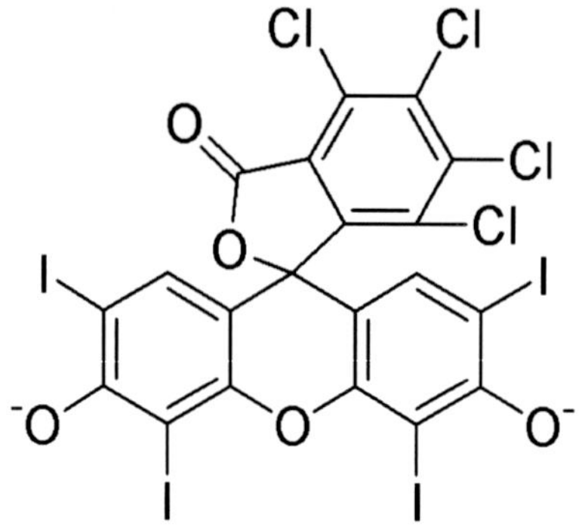	20[b] insulin receptor	7.5 kinesin	16	50	N.D.	80
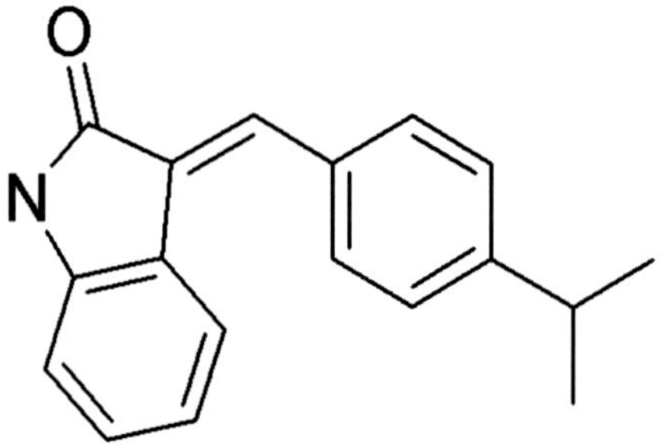	5.2 VEGF	10.0 IGF-1	6	30	30	55
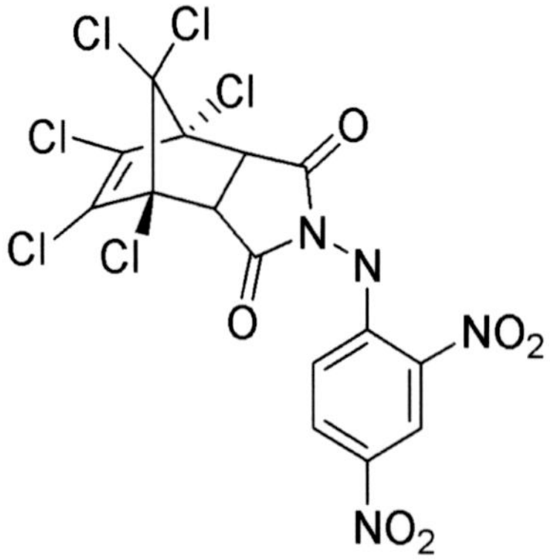	25 farnesyltransferase		3	9	25	150

Table 3, continued. Nonspecific inhibitors discovered by screening (12).

Structure	Original Target(s)		IC$_{50}$ (μM) β-lactamase	Chymotrypsin	cDHFR	β-gal
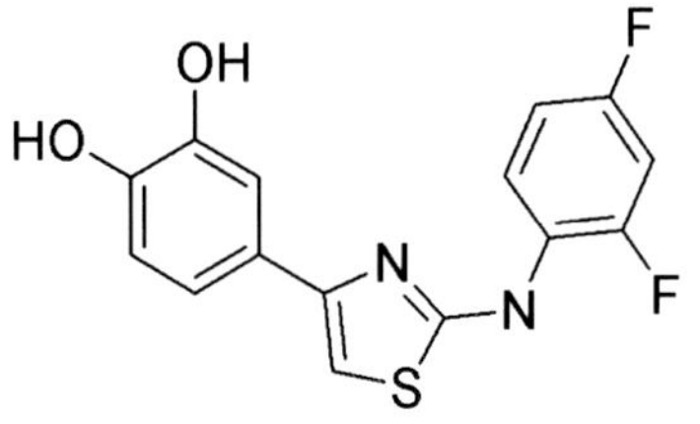 [structure: HO/OH-phenyl thiazole with 2,4-difluorophenylamine]	15[c] gyrase		18	100	150	320
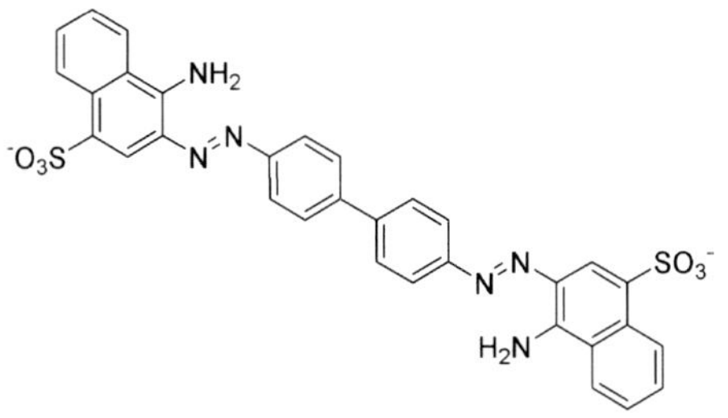 [structure: bis-azo aminonaphthalene sulfonate (Congo red)]	1 prion	30.4 TIM	3.9	40	0.4	100
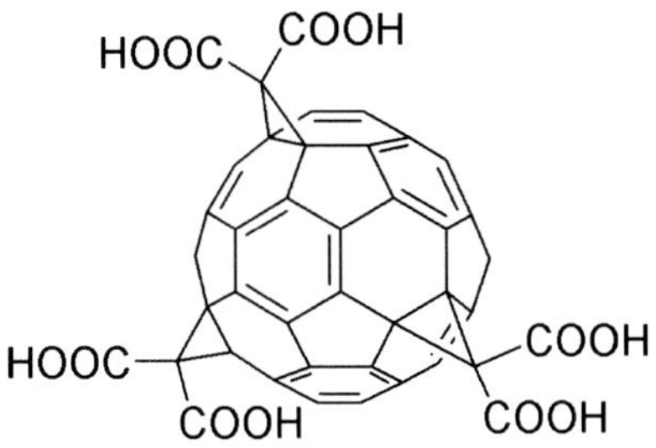 [structure: fullerene tris-malonic acid]	17 eNOS	24 nNOS	7	60	N.D.	N.D.
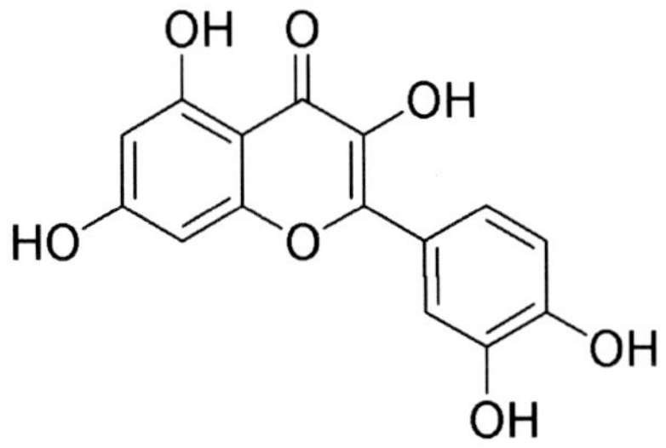 [structure: quercetin]	3.8 P13K	11.0 integrase	4	100	N.D.	220

[a]Our unpublished observations. [b]K_d. [c]maximal non-effective concentration. cDHFR, chicken DHFR; β-gal, β-galactosidase; pDHFR, *Pneumocystis carinii* DHFR; TS, thymidylate synthase; VEGF, vascular endothelial growth factor receptor tyrosine kinase; IGF-1, insulin-like growth factor receptor tyrosine kinase; TIM, triosephosphate isomerase; eNOS, endothelial nitric oxide synthase; nNOS, neuronal nitric oxide synthase; PI3K, phosphoinositide 3-kinase; N.D., not determined

The experiment that put us onto the right way of thinking (after months of befuddlement) was increasing the enzyme concentration ten-fold, while leaving the inhibitor concentration untouched. Since this involved raising β-lactamase from 1 nM to 10 nM, and left the average inhibitor at 10 μM, this should have had no effect on inhibition levels. But instead it attenuated them dramatically. We wondered if the inhibitory species was not a single small molecule, or even two or three, but an aggregate of thousands.

If an aggregate was responsible for inhibition, it should measurable by direct methods. Using dynamic light scattering (DLS) we found that in common buffers these "inhibitors" formed particles of 50 to 450 nm in diameter-almost two orders of magnitude larger than the enzymes that they inhibited. These aggregates were also observed by transmission electron microscopy (TEM). These results are consistent with the hypothesis that these promiscuous inhibitors are acting by forming an aggregate in solution, and that it is these aggregates that inhibit enzymes non-specifically.

In a final experiment, we turned to compounds from the Pharmacia screening database, and asked whether promiscuous, aggregating inhibitors occurred among them. Of the thirty compounds we investigated, twenty were promiscuous, aggregate-forming inhibitors.

In summary, we propose that a single mechanism of action underlies the inhibition pattern of many non-specific inhibitors that have been, and still are being, discovered by virtual and high throughput screening. A burning question to many is how one might recognize such inhibitors in advance, using chemical similarity techniques. This is a question that we cannot at this time answer - the compounds that show this behavior are only very loosely similar, and there are exceptions to every rule we have considered. What is clear is that there are unambiguous experimental tests that can identify such aggregating inhibitors. Such diagnostic experiments should be routinely performed before carrying forward a discovery project.

Reprise: Hits, Leads and Artifacts from Docking and High Throughput Screening

We return to the question posed at the beginning of this essay: why do docking at all, why not just screen experimentally? In well-controlled cases, docking can propose sensible novel ligands and can do so with some accuracy.

The cavity sites in lysozyme provide model systems for testing developments in docking programs, our own and those of others. Although the right head-to-head comparison between docking and HTS has yet to be performed (in PTP1B we used different databases), the experience with PTP1B (3) and with several other systems (13) suggests that structure based efforts in discovery may be considered as alternatives to HTS.

Among the largest challenges facing both docking and HTS is that of promiscuity through aggregation. Small molecules have the option not only of binding to a receptor, but also of aggregating together. Such aggregates inhibit many enzymes non-specifically. In addressing this problem, docking and HTS are allies. Both techniques will gain much from eliminating these promiscuous inhibitors from their hit-lists (14, 15). An encouraging aspect to emerge from these early studies is that there are clear diagnostic tests for these inhibitors. These will allow investigators to eliminate aggregating inhibitors early and thereafter to focus on the true ligands that emerge from structure-based methods, which hold such promise for lead discovery.

REFERENCES

[1] Abagyan, R. & Totrov, M. (2001). *Curr. Opin. Chem. Biol.* **5**:375-82.

[2] van Gunsteren, W. F. & Berendsen, H. J. C. (1990). *Angew. Chem. Int. Ed. Engl.*, **29**:992-1023.

[3] Doman, T. N. et al. (2002). *J. Med. Chem.*,in press.

[4] Eriksson, A. E., Baase, W. A., Wosniak, J. A., Matthews, B. W. (1992). *Nature* **355**:371-373.

[5] Morton, A. & Matthews, B. W. (1995). *Biochemistry* **34**:8576-8588.

[6] Su, A. I. et al. (2001). *Proteins* **42**:279-293.

[7] Lorber, D. M. & Shoichet, B. K. (1998). *Protein Sci.* **7**:938-950.

[8] Shoichet, B. K., Leach, A. R., Kuntz, I. D. (1999). *Proteins* **34**:4-16.

[9] Lorber, D. M., Udo, M. K., Shoichet, B. K. (2002). *Protein Science*, in press.

[10] Li, J. B., Zhu, T. H., Cramer, C. J., Truhlar, D. G. (1998). *Journal of Physical Chemistry A* **102**:1820-1831.

[11] Puius, Y. A. et al. (1997). *Proc. Natl. Acad. Sci. USA.* **94**:13420-5.

[12] McGovern, S. L., Caselli, E., Grigorieff, N., Shoichet, B. K. (2002). *J. Med. Chem.* **45**:1712-1722.

[13] Paiva, A. M. et al. (2001). *Biochim. Biophys. Acta* **1545**:67-77.

[14] Rishton, G. M. (1997). *Drug Discov. Today* **2**:382-384.

[15] Roche, O. et al. (2002). *J. Med. Chem.* **45**:137-142.

HIGH-THROUGHPUT X-RAY TECHNIQUES AND DRUG DISCOVERY

HARREN JHOTI

Astex Technology Ltd, 250 Cambridge Science Park, Cambridge CB4 0WE, UK

E-Mail: h.jhoti@astex-technology.com

Received: 18th June 2002 / Published: 15th May 2003

BACKGROUND

In the past two decades the promise of structure-based drug design has continued to attract significant interest from the pharmaceutical industry. The initial wave of enthusiasm in the late eighties resulted in some notable successes, for example, the crystal structures of HIV protease and influenza neuraminidase were used to design Viracept and Relenza, both drugs currently used in anti-viral therapy (1, 2). However, although structure-based design methods continued to be developed, the approach became largely eclipsed in the early nineties by other technologies such as combinatorial chemistry and high-throughput screening (HTS) which seemed to offer a more effective approach for drug discovery. The goal of obtaining a crystal structure of the target protein, particularly in complex with lead compounds was regarded as a resource-intensive, unpredictable and slow process. During that period it was clear that protein crystallography was unable to keep pace with the other drug discovery technologies being performed in a high-throughput mode.

More recently, there has been resurgence in interest for using structure-based approaches driven largely by major technology developments in protein crystallography that have resulted in crystal structures for many of today's therapeutic targets. Furthermore, the ability to rapidly obtain crystal structures of a target protein in complex with small molecules is driving a new wave of structure-based drug design. In this chapter I will briefly describe some of these technology developments and focus on how they have enabled high-throughput X-ray crystallography to be applied to drug discovery.

Technology Advances

There are many areas in which new technologies and methods are being developed to enable high-throughput structure determination by X-ray crystallography (3, 4). The process from gene to crystal structure is clearly multidisciplinary and advances in molecular biology, biochemistry, crystallisation, X-ray data collection and computational analysis underpin high-throughput X-ray crystallography. Many of these advances are being made in the public-initiatives focused on structural genomics. The most progressed and well-funded initiatives are found in the US where the NIGMS (National Institute of General Medical Sciences) is planning to spend US$ 150M and is currently funding nine structural genomics centres under its Protein Structure Inititiative (5). Similar programs are underway in other countries, for example, the Protein Structure Factory in Germany is focusing on solving structures of human proteins in collaboration with the German Human Genome Project (DHGP) and the Japanese government is supporting the RIKEN Structural Genomics Initiative.

The main focus of these structural genomics initiatives is to automate all steps of the protein crystallographic process and apply the methods to determine structures of proteins for which no three-dimensional information exists (6). In addition to these publicly-funded centres, some specialist biotechnology companies have also been formed to pursue structural genomics programs. These include Structural GenomiX and Syrrx, both based in San Diego (US), who are developing significant automation to streamline the gene to crystal structure process (7).

Clone to Crystal

Expression, purification and characterisation of a novel protein in a quantity and form that is suitable for crystallisation and X-ray analysis probably occupies over 80% of the time in most structural biology groups. Consequently, methods for high-throughput parallel expression and purification are now being developed in many laboratories (8). Typically, 10-50 mgs of protein is required to screen sufficient numbers of crystallisation conditions to obtain initial crystals. Traditionally, a handful of different DNA constructs would be generated, after analysis of the protein sequence, in an attempt to remove flexible regions of the protein that may hinder crystallisation. Each construct would then be tested for expression in the host cell, usually *Escherichia coli* or insect cells, and the level of functional protein analysed using bioassay and polyacrylamide gel electrophoresis (PAGE).

In the past these different constructs would be analysed sequentially, but recent developments in molecular biology, based on DNA recombination, now enable high-throughput approaches for cloning and expression where tens to hundreds of DNA constructs can be easily generated to test in parallel for high expression. Protein purification has also seen significant improvements owing to the development of affinity tags that allow proteins to be purified significantly faster and more efficiently (9). Automated methods based on affinity chromatography, such as a nickel-nitrilotriacetic acid (Ni-NTA) column, are now available which can process samples in parallel using a 96-well format.

Crystallisation is often regarded as a slow, resource-intensive step with low success rates in obtaining good quality crystals. However, much of the failure during this step can be attributed to poor quality protein samples that often have some level of chemical or conformational heterogeneity.

The use of biophysical methods, such as dynamic light scattering, to rigorously characterise the protein sample is a key step before performing crystallisation experiments. Significant advances in automation have also improved the process of crystallisation with the new generation of robots able to efficiently sample the multidimensional space by varying precipitant concentration, buffers and pH - all variables known to affect crystallisation. Video systems are being developed that allow the user to monitor the crystallisation experiment using image recognition techniques (10).

CRYSTAL TO STRUCTURE

Once X-ray quality crystals have been grown, data collection using several wavelengths or derivatives is required in order to obtain the protein structure. X-ray data collection has been revolutionised in the last decade by both better X-ray sources and detectors. Third generation synchrotrons are now available across the world which provide high intensity X-ray beams allowing the data collection time to be significantly reduced (11). Synchrotron radiation coupled with charged-coupled device (CCD) detectors have allowed complete X-ray datasets for a crystal to be collected and processed within hours instead of days. High-throughput X-ray data collection has required the development of robotic systems that store and mount crystals sequentially while maintaining the samples at liquid-nitrogen temperatures (12, 13).

Phase determination has also become dramatically easier by the application of synchrotron radiation to single and multi-wavelength anomalous diffraction techniques, known as SAD and

MAD, respectively. Finally, new methods of electron density interpretation and model-building have allowed rapid and automated construction of protein models without the need for significant manual intervention (14).

Structure-Based Lead Discovery

All these technology advances have resulted in an exponential increase in the number of crystal structures being deposited into the Protein Data Bank (PDB) in recent years (15). Currently, the PDB holds nearly 18,000 protein structures, most of which have been determined using X-ray crystallography (Fig1).

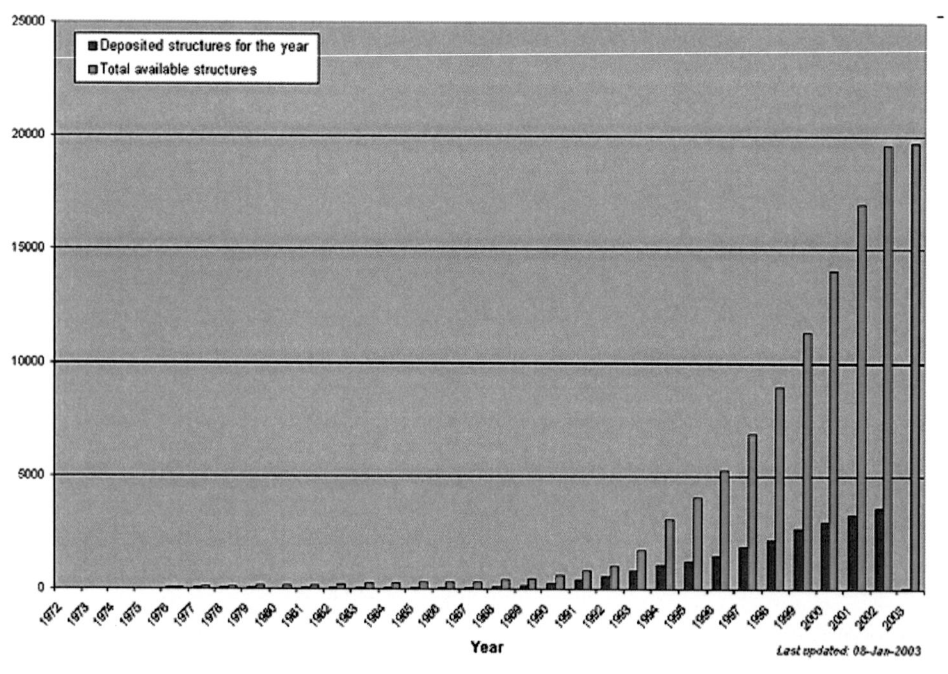

Figure 1. Growth in the Protein Data Bank. For many years the number of protein structures being determined and deposited into the PDB was linear, however, with the advent of major technology advances over the last decade the deposition rate has become exponential. (Source: The Protein Data Bank at www.rcsb.org; Berman *et al. Nucleic Acids Research*, **28** 235-242, 2000).

Due to this growing wealth of protein structure data, it is increasingly likely that the three-dimensional structure of a therapeutic target of interest to drug discovery scientists will already have been determined. Furthermore, it is expected that within the next five years, crystal structures of a large majority of the non-membrane protein targets of interest to the pharmaceutical industry will be available.

Although the structure of the native target protein is a useful start to guide a lead discovery program, the maximum value is derived only from structures of the protein in complex to potential lead compounds. This is due to the fact that many proteins undergo some level of conformational movement on ligand binding which has proved very difficult to predict from the native structure alone. Furthermore, water molecules often play a key role in the interactions between small molecules and proteins and their positions need to be established experimentally. The ability to rapidly determine crystal structures of protein-ligand complexes is required to effectively guide the lead optimisation phase, but may also allow X-ray crystallography to be applied to drug discovery in a new way: as a screening tool (4).

The most reliable approach to determine the structure of a protein-ligand complex, is either by co-crystallisation or by soaking the ligand into the preformed crystal. However, when X-ray crystallography is used as a method for ligand screening, the soaking option is much preferred. After collecting the X-ray data from a protein crystal exposed to a ligand, the next step is to analyse and interpret the resulting electron density. This step is often time consuming and requires a crystallographer to spend several days assessing the data from a single protein/ligand experiment. This is a key bottleneck for the use of X-ray crystallography as a method for screening compounds. Technology advances have now been made to automate and accelerate this step. Software tools such as Quanta from Accelrys Inc. (San Diego, CA, USA) and AutoSolve® from Astex (Cambridge, UK) can assist the crystallographer in the analysis and interpretation steps.

FRAGMENT-BASED LEAD DISCOVERY

There is growing interest in the use of molecular fragments for lead discovery. One reason for this interest is due to a problem that is evident in the nature of 'hits' identified from traditional bioassay-based High Throughput Screens (HTS). The average MW of successful drugs in the World Drug Index is in the low 300s, which is similar to the average MW in current corporate collections (16). This implies that corporate compound collections have evolved to be broadly "drug like" with respect to MW and other features. However, recent publications conclude that hits from a HTS should have a lower molecular weight than drugs, that is screening drug-like compounds may not be the most effective way to find good lead compounds (17). This conclusion is based on the expected increase in molecular weight, of about 80, during the lead optimisation process. Therefore, a HTS hit from a corporate compound collection with μM affinity towards the target may well already have an "average drug MW" yet it is likely that the

MW will increase very significantly during the lead optimisation process, leading to significantly poorer drug like properties with respect to solubility, absorption and clearance (18).

In order to address this issue several groups have been developing methods to identify low MW fragments (MW 100-250) that could be efficiently optimised into novel lead compounds possessing good drug like properties. These molecular fragments would by definition have limited functionality and would therefore exhibit weaker affinity (typically in the 50 μm-mM range). This affinity range is outside of the normal HTS sensitivity range and as such cannot routinely be identified in standard bioassays due to the high concentration of compound that would be required, interfering with the assay and leading to significant false positives. Rather than trying to push bio-assays into this affinity range, people are turning increasingly to biophysical methods such as NMR and X-ray crystallography for fragment-based screening approaches. For example, Fesik and colleagues have pioneered methods in which NMR is used to screen libraries of molecular fragments (19, 20). In determining structure-activity relationships (SAR) by NMR, perturbations to the NMR spectra of a protein are used to indicate that ligand binding is taking place and to give some indication of the location of the binding site. Once molecular fragments bound to the target protein have been identified they can then by linked together or 'grown' using structure-based chemical synthesis to improve the affinity for the target protein (Fig. 2).

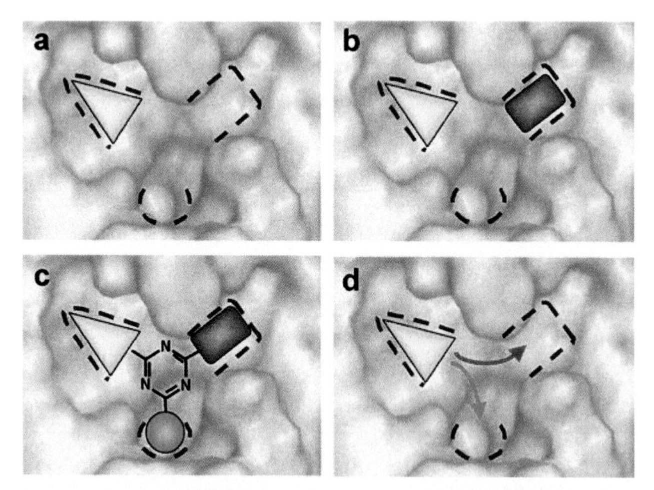

Figure 2. Once fragments have been identified bound into the active site they can be used as a start-point for iterative structure-driven chemistry resulting in a drug-size lead compound. If two fragments are bound in two different pockets (b) they could be used to decorate an appropriate scaffold (c). Alternatively, a single fragment could be rationally modified to occupy other neighbouring pockets (d).

Fragment-Based Screening Using X-ray Crystallography

X-ray crystallography has the advantage of defining the ligand-binding sites with more certainty than NMR and the binding orientations of the molecular fragments play a critical role in guiding efficient lead optimisation programs. Different sets of molecular fragments can be used to target a particular protein.

For example, in a screen of fragments against trypsin, a 'focused set' was selected based on known binders such as benzamidine, 4-aminopyridine and cyclohexylamine (21). These molecules were each used as starting points for similarity searches of chemical databases. Representatives from these searches were then purchased or synthesised and dissolved in an organic solvent (such as dimethylsulphoxide (DMSO)) added to a single protein crystal, and then left to soak for 1 hour to give the molecule time to penetrate into the active site.

The concentration of the molecular fragment is typically greater than 20 mM, reflecting the low-affinity that is expected. Fragment libraries can be screened as singlets or in cocktails using X-ray crystallography. As the output from an X-ray experiment is a visual description of the bound compound (its electron density) it is possible to screen cocktails of compounds without the need to deconvolute. An optimum cocktail size is typically between 4-8 and is defined by the tolerance of the protein crystals to organic solvents and the concentration at which you wish to screen each fragment. For example, if the maximum tolerated solvent concentration is 240 mM then you can screen 8 compounds each at a concentration of 30 mM.

Some of the first experiments in which X-ray crystallography was used as a 'screening tool' were reported by Verlinde and colleagues who exposed crystals of trypanosomal Triosephosphate Isomerase to cocktails of compounds in their search for inhibitors (22). More recently, Greer and colleagues have described a method for screening using X-ray crystallography that focuses on soaking the target crystals with cocktails of compounds having differing shapes that can easily be distinguished by visual inspection of electron density (23). However, to fully exploit X-ray crystallography as a screening approach it is desirable to implement an objective and automated process to address the key bottleneck of data interpretation and analysis (4). AutoSolve® allows rapid and automated analysis of electron density from fragment soaking experiments using singlets and cocktails of compounds. Examples of electron density that were unambiguously interpreted by AutoSolve® are shown in Fig 3.

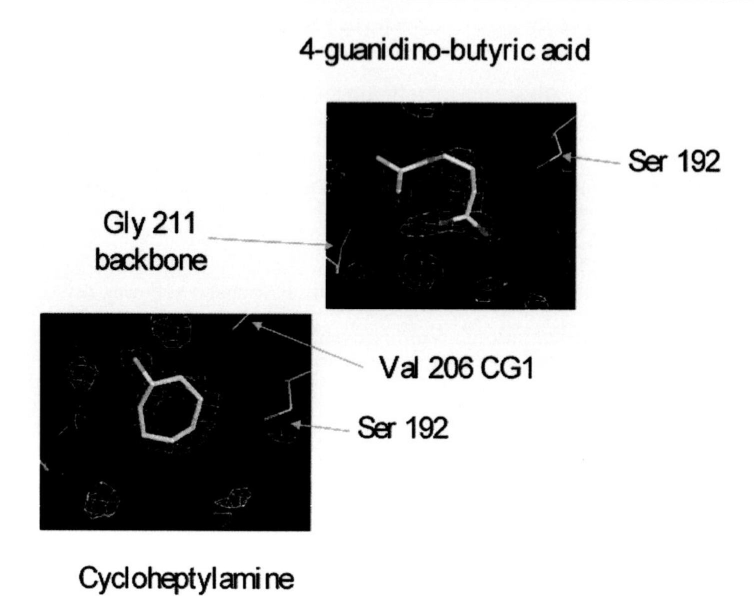

Figure 3. AutoSolve® interpretation of single compounds. Electron density can be automatically interpreted for small weak-binding fragments using AutoSolve®. Although the binding affinity is weak (IC_{50} = 1 mM for cyclohexylamine) the interactions with the protein are clearly defined.

In each case the binding mode of the small-molecule fragment is clearly defined by the electron density, which means that although the affinity may be in the millimolar range, the binding is ordered with key interactions being made between the compound and the protein. In fact, AutoSolve® requires no human intervention if the quality of electron density is high, and can identify the correct compound bound at the active site from an experiment where the crystal has been exposed to a cocktail of compounds (Fig 4).

Another key advantage of using molecular fragments for screening is the significant amount of chemical space that is sampled using a relatively small library of compounds. For example, if the binding of several heterocycles is probed against specific binding pockets in a protein, the discrimination between a binding and non-binding event depends solely on the molecular complementarity and is not constrained or modulated by the heterocycle being part of a larger molecule. This is a far more comprehensive and elegant way to probe for new interactions than having the fragments attached to a rigid template, as might derive from a conventional combinatorial chemistry approach.

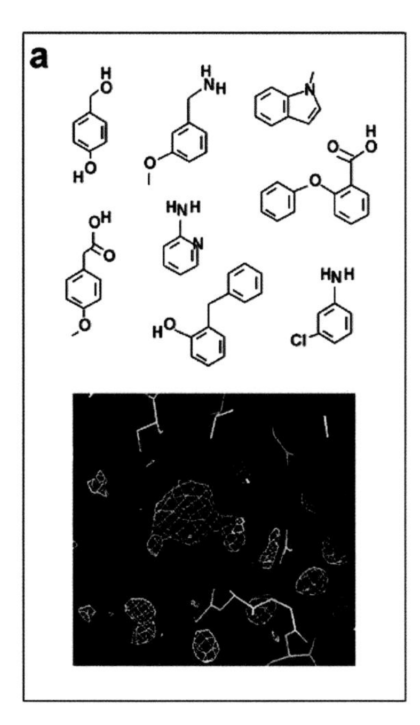

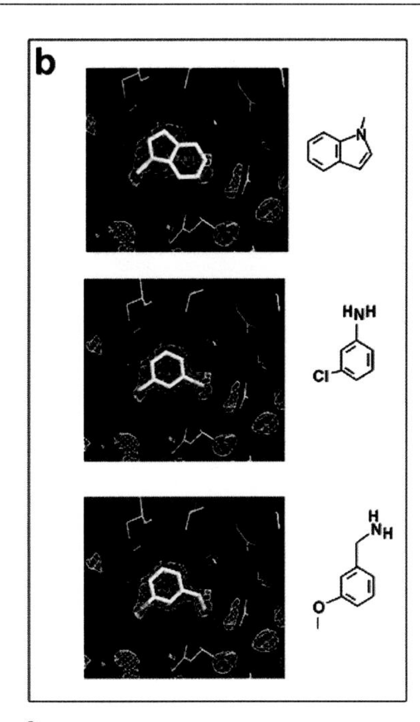

Figure 4. Analysing fragment cocktails using AutoSolve® A crystal was exposed to a cocktail of 8 fragments and the reultant electron density is shown (**A**). Each of the eight molecules is fitted into the electron density by AutoSolve® and the optimal fit is identified by the program (**B**).

STRUCTURE-BASED LEAD OPTIMISATION

Determination of the binding of one or more molecular fragments in the protein active site provides a starting point for medicinal chemistry to optimise the interactions using a structure-based approach. The fragments can be combined onto a template or used as the starting point for 'growing out' an inhibitor into other pockets of the protein (Fig. 2). The potency of the original weakly-binding fragment can be rapidly improved using iterative structure-based chemical synthesis. For example, in one of our lead discovery programs targeted against p38 kinase, we identified an initial fragment, AT464 (MW=X), which exhibited an IC_{50} of 1 mM in an enzyme assay.

Using the crystal structure of AT464 bound to the protein kinase we were able to improve potency more than 20-fold by synthesising only 20 analogues. The resulting compound, AT660, had an IC_{50} of 40 μM (unpublished results). Compounds from this novel lead series were further optimized to improve potency using rapid structure-based chemical synthesis. This resulted in the current lead compound, AT1731, which has an IC_{50} of 100 nM against the enzyme and is active in inhibiting TNF release in LPS-stimulated cells. This improvement in affinity is produced by iteratively increasing the number of interactions between the protein and the compound (Fig. 5).

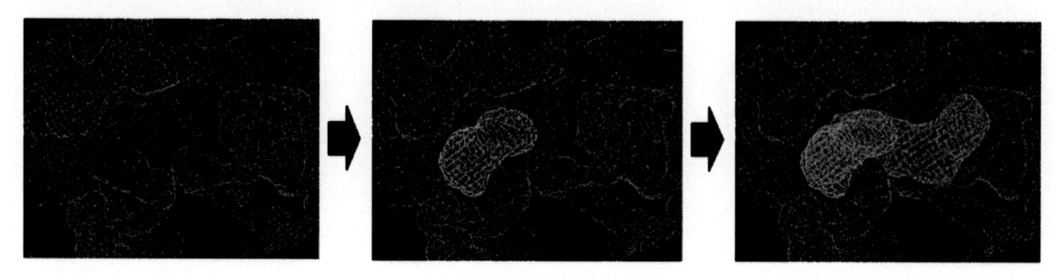

Figure 5. Optimisation of initial low affinity fragment into potent lead compound. The initial molecular fragment is used as a starting point from which extra protein/ligand interactions are built, guided by the 3-D structure of the protein. This can be seen in the increasing volume of occupation within the protein active site.

Using such a structure-based chemistry strategy, progressing from millimolar hits to nanomolar leads for our first lead series required the synthesis of <250 compounds. More recently, we have identified a second lead series for p38 kinase with a structurally distinct template, again by optimising a weakly-binding molecular fragment using structure-based synthesis.

CONCLUSIONS

The role of protein structure within the drug discovery process is likely to increase significantly over the coming years as more and more crystal structures become available for the therapeutic targets. This will no doubt fuel an increase in structure-based drug design programs which look to optimise lead compounds that were initially identified using traditional HTS campaigns. Recent technology advances in structure determination may also allow X-ray crystallography to be used as a method for ligand screening. This may have particular value for fragment-based lead discovery where the initial molecular fragments are likely to have an affinity too weak to enable detection using traditional bioassay-based methods. Initial data generated using X-ray crystallographic screening of molecular fragment libraries indicates that novel scaffolds can be identified and subsequently optimised using rapid structure-based synthesis to generate useful lead compounds. The potential of this fragment-based screening approach using X-ray crystallography may be significant, particularly against targets which have remained intractable using conventional screening methods.

ACKNOWLEDGEMENTS.

I wish to thank Drs. Mike Hartshorn and Ian Tickle who developed AutoSolve® and Dr. Robin Carr for useful discussions and for reviewing the manuscript. I also appreciate the assistance of Dr. Emma Southern in the production of this manuscript.

This manuscript first published in: Ernst Schering Research Foundation Workshop, Series Volume 42: Waldmann/Koppitz: Small Molecule Protein Interaction, Springer Verlag 2003

REFERENCES

[1] Kaldor S. W. et al. (1997). Viracept (Nelfinavir Mesylate, AG1343): A potent, orally bioavailable inhibitor of HIV-1 protease. *J. Med. Chem.* **40**:3979-3885.

[2] von Itzstein, M. et al. (1993). Rational design of potent sialidase-based inhibitors of influenza virus replication. *Nature* **363**:418-423.

[3] Heinemann U. et al. (2001). High-throughput three-dimensional protein structure determination. *Curr. Opin. Biotech.* **12**: 348-354.

[4] Blundell T. L. et al. (2002). High-throughput crystallography for lead discovery in drug design. *Nat. Rev. Drug Disc.* **1**:45-54.

[5] Norvell J. C. & Machalek A. Z. (2000). Structural genomics programs at the US National Institute of General Medical Sciences. *Nat. Struc. Biol.* **7**:931.

[6] Vitkup D. et al. (2001). Completeness in structural genomics. *Nat. Struct. Biol.* **8**:559-566.

[7] Dry S. et al. (2000). Structural genomics in the biotechnology sector. *Na.t Struc. Biol.* **7**:946-949.

[8] Lesley S. A. (2001). High throughput proteomics: protein expression and purification in the post-genomic world. *Protein Exp. Purif.* **22**:159-164.

[9] Crowe J. et al. (1994). 6xHis-Ni-NTA chromatography as a superior technique in recombinant protein expression/purification. *Methods Mol. Biol.* **31**:371-387.

[10] Stewart L. et al (2002). High-throughput crystallisation and structure determination in drug discovery. *Drug Disc. Today* **7**:187-196.

[11] Hendrickson W. (2000). Synchrotron crystallography. *Trends. Biochem. Sci.* **25**:637-643.

[12] Abola E. et al. (2000). Automation of X-ray crystallography. *Nat. Struc. Biol.* **7**:973-977.

[13] Muchmore S. W. et al. (2000). Automated crystal mounting and data collection in protein crystallography. *Structure* **8**:R243-R246.

[14] Perrakis A. et al. (1999). Automated protein model building combined with iterative structure refinement. *Nat. Struc. Biol.* **6**:458-463.

[15] Berman H. M. (2000). The Protein Data Bank and the challenge of structural genomics. *Nat. Struc. Biol.* **7**:957-959.

[16] Oprea T. I. (2001). Is there a difference between Leads and Drugs? A Historical Perspective. *J. Chem. Inf. Comp. Sci.* **41**:1308-1315.

[17] Hann M. et al. (2001). Molecular complexity and its impact on the probability of finding leads for drug discovery. *J. Chem. Inf. Comp. Sci.* **41**:856-864.

[18] Lipinski C. A. et al. (2001). Experimental and computational approaches to estimate solubility and permeability in drug discovery and development. *Adv. Drug Delivery Rev.* **46**:3-26.

[19] Shuker S. B. et al. (1996). Discovering high-affinity ligands for proteins: SAR by NMR. *Science* **274**:1531-1534.

[20] Hajduk P. J. et al. (1999). NMR-based screening in drug discovery. *Quart. Rev. Biophys*. **32**:211-240.

[21] Blundell T. L. et al. High throughput X-ray crystallography for drug discovery. *Proceedings of the Royal Society of Chemistry* meeting *Cutting Edge Approaches to Drug* Design, March 2001 (Flower, D ed.) RSC Publications Dept, London, (in press).

[22] Verlinde C. et al. (1997). Antitrypanosomiasis drug development based on structures of glycolytic enzymes. Structure-based Drug Design (ed. Veerapandian, P) 365-394 (Marcel Dekker, Inc, New York, NY.

[23] Nienaber V. L. et al. (2000). Discovering novel ligands for macromolecules using X-ray crystallographic screening. *Nat. Biotech.* **18**:1105-1108.

Computer-Aided Decision Making in Pharmaceutical Research

Gerald M. Maggiora

Computer-Aided Drug Discovery, Pharmacia Corporation, 301 Henrietta Street, Kalamazoo, MI 49007-4940, USA

E-Mail: gerald.m.maggiora@pharmacia.com

Received: 7th June 2002 / Published: 15th May 2003

Abstract

A description of a computer-aided decision making methodology, called the Analytic Hierarchy Process (AHP), is presented. The method was developed by Thomas Saaty over three decades ago to handle a variety of business-oriented decision making activities. The AHP is a flexible methodology that allows both subjective and objective data to be considered in a decision process. Moreover, it is intuitive and relatively easy to understand the way in which decisions are made. Although many business-related applications have been carried out over the years, very few science-based applications currently exist. In addition to a description of the basic methodology an example from drug-discovery research, namely biological target selection, will be presented as an illustration of how the AHP methodology can be applied in pharmaceutical research. A brief mention of other possible applications will also be provided.

Introduction

Decision making methodologies have been applied in a broad range of situations for many years. Most applications to date have been in business-related activities. This is necessitated by the number and complexity of the issues that bear upon many business decisions. Significant advances in computer software and hardware have also played a major role by providing the "computer power" necessary to treat decision problems more realistically. In pharmaceutical research, especially in large pharmaceutical companies where many projects are going on simultaneously, many of the same types of decision problems exist. However, in contrast to other business areas, decision theoretic approaches are essentially non-existent. One of the reasons for this may be the perceived difficulty of properly formulating research-based decision problems, which involve both *quantitative* and *qualitative* variables. Moreover, the reasoning behind decision-theoretic methodologies and the results obtained from them are often non-intuitive and difficult to understand.

In the seventies Thomas Saaty developed a decision-theoretic methodology, called the *Analytic Hierarchy Process*, that is relatively simple conceptually and thus, may be more suitable to research-based decision problems. Details of his methodology were described in his first book (1). The AHP represents a fundamental approach that is based upon pairwise comparisons, is designed to cope with both the *rational* and *intuitive* aspects of a decision problem, and is capable of selecting the best alternatives with respect to a number of competing criteria. Importantly, the AHP allows for inconsistencies in judge-ments and affords a means for improving consistency. Table 1 provides a brief listing of the some the types of decision problems that the AHP has been applied to. A number of books by Saaty and others (2,3,4,5) describe numerous types of applications with examples. More recently Saaty has generalised his theory to deal with dependence and feedback (6). Interestingly, very few applications in chemical and biological research have appeared.

Table 1. A sample of the breadth of AHP applications.

Architectural Design	Technological Choices
Conflict Resolution	Marketing Strategies
Performance Evaluations	Pricing Strategies
Student Admissions	Environmental Decisions
I/O Analysis	Cost-Benefit Analysis
Economic Forecasting	Transportation Systems
Oil Prospecting	Musical Compositions
Selection of Bridge Type	Movie Criticism

The AHP, as its name implies, deals with decision problems that can be structured hierarchically. Figure 1 depicts a simple three-level hierarchy. As is seen from the figure, the ‘Goal’ is evaluated with respect to the three ‘Criteria’ that each subsume the entire set of ‘Alternatives.’ The relative importance or ranking of each criterion to the decision goal is determined from pairwise comparisons among the criteria. Pairwise comparisons are based upon *relative* measurements that characterise the ‘dominance’ of one criterion with respect to another. As it is used here, ‘dominance’ is taken as a generic term that characterises the dominance, importance, desirability, likelihood, or whatever term is appropriate, of one criterion over another.

Many of the criteria dealt with in type of decision problems illustrated in this work are *intangible* and hence, their relative measurements are largely subjective.

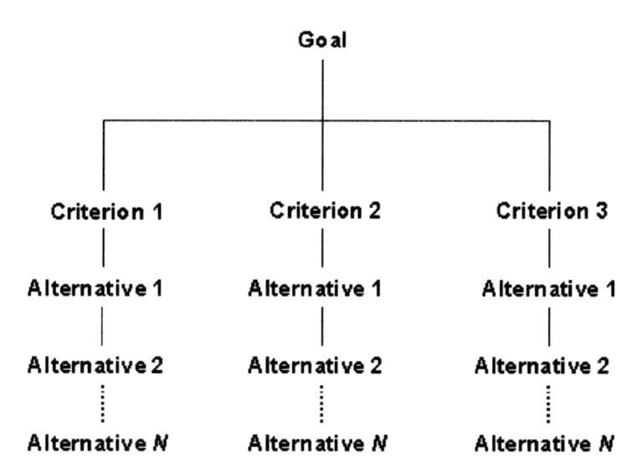

Figure 1. Example of a simple hierarchy consisting of a goal, three subordinate criteria relevant to the goal, and the N alternatives with respect to each of the criteria.

For example, in selecting a target for drug discovery in a large pharmaceutical company (*vide infra*), how important is 'Unmet Medical Need' compared to the company's 'Intellectual Property' with respect to the target?

While this may seem a bit like comparing 'apples' to 'oranges', it is something that humans do, subjectively, all of the time. Psychologists have studied such comparative assessments for many years and have determined that humans can only effectively handle about nine levels or gradations in making subjective, comparative assessments (7), as summarised in Table 2.

Table 2. The Fundamental Scale (7).

Intensity of Importance	Definition	Explanation
1	Equal Importance	Two activities contribute equally to the objective
2	Weak	
3	Moderate Importance	Experience and judgement slightly favour one activity over another

In addition, a *reciprocal relationship* exists such that if, for example, bioactivity is deemed to be twice as important a criterion as, say, solubility, then solubility must be only half as important as bioactivity. All of the pairwise comparisons among the criteria are elements of the pairwise comparison matrix or simply the *comparison matrix* (see *e.g.* Eq. (1)). The values of the components of the *principal eigenvector* of the comparison matrix are all positive and correspond, with suitable normalisation, to the relative ranking of the criteria, which sum to unity. Thus, the relative ranking is a linear order that is generated from a set of pairwise comparisons.

The decision 'Alternatives,' on the other hand, are ranked with respect to each criterion using an absolute measurement scale appropriate to that criterion. For example, 'very high,' 'high,' 'medium,' 'low,' and 'very low' represent a possible scale, which could be given values 4, 3, 2, 1, 0, respectively. As has been noted by many cognitive psychologists this is well within the range of nine levels that humans can effectively discriminate (1,2,3). The final decision is achieved by weighting the result obtained for a given alternative by the relative ranking of the corresponding criterion and then summing over the three criteria. Each alternative is then placed in an ordered list with respect to its overall "score". Importantly, computing the score for a new alternative can be carried out independently of all other previously scored alternatives, which is a significant benefit when large numbers of alternatives are being considered as illustrated by the example described in this work. In many applications, alternatives are treated in an analogous fashion to criteria (*vide supra*), that is the alternatives are directly compared to each other and not to an absolute scale, but such comparisons are inappropriate in most of the types of research applications of AHP considered here. This is because in an absolute scale each alternative is evaluated separately. Adding a new alternative does not influence the values associated with any of the alternatives considered previously, and does not change their rankings relative to on another. However, the new alternative can, depending upon its value, be inserted anywhere in the previously ranked list of alternatives. This is quite advantageous in many of the types of situations in pharmaceutical research where computer-aided decision making may play a role.

The basic methodology will be presented in the METHODOLOGY section, followed in the RESULTS AND DISCUSSION section by an example based on selecting a "biological target" for drug discovery. The final section - CONCLUSIONS AND FUTURE WORK - provides a summary of the material and draws several conclusions regarding the applicability of the AHP approach to decision making in pharmaceutical research. All of the work presented here was carried out with the software product *EXPERTCHOICE2000*™ (8).

METHODOLOGY

As has been noted above, pairwise comparison is a key element of AHP methodology. A comparison matrix, **A,** is used to determine the relative dominance, order, importance, priority, likelihood, etc. among a set of n criteria $\{C_1, C_2, ..., C_n\}$. Each element of **A**, $a_{i,j}$, is obtained by comparing criteria according to an appropriate scale: $a_{i,j}$ corresponds to how much the i-th criterion is 'favoured' over the j-th criterion. Because of the reciprocal property of these comparative judgements $a_{i,j}=1/a_{j,i}$ so, for example if $a_{i,j}=3$, then $a_{j,i}={}^1/_3$. The comparison matrix is a positive, reciprocal matrix and is as shown in eigenvalue form in Eq. (1)

$$\begin{bmatrix} 1 & a_{1,2} & \cdots & a_{1,x} \\ a_{1,2}^{-1} & 1 & \cdots & a_{2,x} \\ \vdots & \vdots & \ddots & \vdots \\ a_{1,x}^{-1} & a_{2,x}^{-1} & \cdots & 1 \end{bmatrix} * \begin{bmatrix} w_1 \\ w_2 \\ \vdots \\ w_x \end{bmatrix} = \lambda_{\max} * \begin{bmatrix} w_1 \\ w_2 \\ \vdots \\ w_x \end{bmatrix} \quad (1)$$

where $\lambda_{\max}$ is the principal eigenvalue, $[w_1, w_2, ..., w_n]^T$ the principal eigenvector, and 'T' represents the transpose. Because **A** is a *positive*, *reciprocal matrix* the components of its principal eigenvector are all positive (1,2,4,6) and in this work are normalised in either of two ways:

$$\bar{w}_i = \frac{w_i}{\sum_{j=1}^{n} w_j} \Rightarrow \sum_{j=1}^{n} \bar{w}_j = 1 \quad (2a)$$

or

$$\hat{w}_i = \frac{w_i}{\max(w_1, w_2, ..., w_n)} \Rightarrow \text{where } \hat{w}_i = 1 \text{ if } w_i = \max(w_1, w_2, ..., w_n) \quad (2b)$$

The normalised weights correspond to the relative dominance, importance, priority, likelihood, etc. of each criterion.

An important issue with respect to the comparison matrix is its *reciprocal consistency*, which involves the reciprocal relationship: if $a_{i,j}>1$, then $a_{j,i}<1$. In words, if i-th criterion dominates the j-th criterion, then the j-th criterion cannot also dominate the i-th criterion. This type of consistency is simple to enforce. A more complex form of consistency is *transitive consistency*, namely that $a_{i,j} \cdot a_{j,k}=a_{i,k}$. Again in words, if the i-th criterion dominates the j-th criterion by a factor of, say three, and the j-th criterion dominates the k-th criterion by a factor of, say one-half, then for transitive consistency the i-th criterion must dominate the k-th criterion by a factor of $3 \cdot {}^1/_2={}^3/_2$. Transitive consistency is the most difficult to achieve in practice but can be approached by a careful analysis of the comparative judgements made. As will be seen below, the inconsistency index, I, provides a useful measure of transitive consistency.

Taking the unnormalised components of the principal eigenvector form an 'adjusted' comparison matrix, **A′**, using their ratios. Thus each element of **A′** is in this case given by $a'_{i,j}=w_i/w_j$. A little algebra shows that the principal eigenvector of the 'adjusted' comparison matrix is identical to that of the original comparison matrix and that the eigenvalue is equal to the number of criteria n, as shown in Eq. (3).

$$\begin{bmatrix} \frac{w_1}{w_1} & \frac{w_2}{w_2} & \cdots & \frac{w_1}{w_n} \\ \frac{w_2}{w_1} & \frac{w_2}{w_2} & \cdots & \frac{w_2}{w_n} \\ \vdots & \vdots & \ddots & \vdots \\ \frac{w_n}{w_1} & \frac{w_n}{w_2} & \cdots & \frac{w_n}{w_n} \end{bmatrix} * \begin{bmatrix} w_1 \\ w_2 \\ \vdots \\ w_n \end{bmatrix} = n * \begin{bmatrix} w_1 \\ w_2 \\ \vdots \\ w_n \end{bmatrix} \quad (3)$$

It can be shown (2,6) that $\lambda_{max} \geq n$, so that as $\mathbf{A} \rightarrow \mathbf{A}'$, that is as **A** becomes more transitive consistent, $\lambda_{max} \rightarrow n$. Thus, one measure of consistency is

$$\mathrm{I} = \frac{\lambda_{\max} - n}{n-1}, \text{ where I } \geq 0 \quad (4)$$

which is somewhat reminiscent in form to sample variance.

Consider the set of alternatives $\{A_1, A_2, ..., A_n\}$ and the matrix **R** of alternatives ranked with respect to each of the n criteria $C_1, C_2, ..., C_n$:

$$\mathrm{R} = \begin{bmatrix} A_1(C_1) & A_1(C_2) & \cdots & A_1(C_n) \\ A_2(C_1) & A_2(C_2) & \cdots & A_2(C_n) \\ \vdots & \vdots & \ddots & \vdots \\ A_m(C_1) & A_m(C_2) & \cdots & A_m(C_n) \end{bmatrix} \quad (5)$$

Ranking the alternatives with respect to the overall goal is obtained by weighting a given alternative by each criterion, *i.e.*, $w(C_k)$, and summing the result, which gives a linear form for the i-th alternative

$$A_i(G) = \sum_{k=1}^{n} w(C_k) \cdot A_i(C_k), \; i = 1,2,...,m \quad (6)$$

Alternatively, Eq. (6) can be written in matrix form:

$$\begin{bmatrix} A_1(C_1) & A_1(C_2) & \cdots & A_1(C_n) \\ A_2(C_1) & A_2(C_2) & \cdots & A_2(C_n) \\ \vdots & \vdots & \ddots & \vdots \\ A_m(C_1) & A_m(C_2) & \cdots & A_m(C_n) \end{bmatrix} * \begin{bmatrix} w(C_1) \\ w(C_2) \\ \vdots \\ w(C_n) \end{bmatrix} = \begin{bmatrix} A_1(G) \\ A_2(G) \\ \vdots \\ A_m(G) \end{bmatrix} \quad (7)$$

In the more general case of a multi-level hierarchy with numerous, 'nested' criteria, a multi-linear form results rather than the linear form given in Eq. (6) (2,6). The APPENDIX should be consulted for more details.

RESULTS AND DISCUSSION

Pharmaceutical research spans a wide range of activities from the initial selection of an appropriate drug target, to the identification and optimisation of a set of lead compounds, to studies of drug absorption, distribution, metabolism, excretion, and toxicity, usually called ADMET, to the various clinical phases. In principle, the AHP can be applied throughout this process, although such applications are extremely rare and are non-existent in drug discovery. The following provides a concrete example of how the AHP can be applied in drug discovery to target selection. As is seen in Figure 2, numerous decision subcriteria are grouped under the two main classes of decision criteria, namely 'Business Issues' and Scientific Issues.

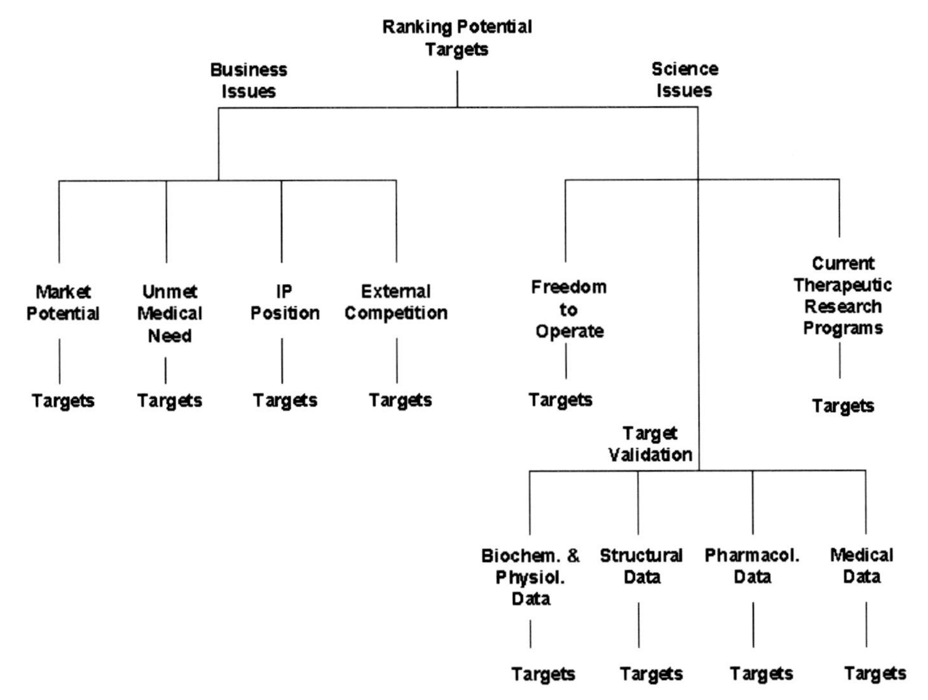

Figure 2. Hierarchy for ranking the suitability of biological targets (*e.g.*, enzymes, receptors,...). Note that "Targets" refers to the total set of targets considered, nine in the case examined in this work.

Business Issues are concerned with four major factors, Market Potential, Unmet Medical Need, Intellectual Property Position, and External Competition. As is clear from Figure 2 Scientific

Issues, namely Freedom to Operate, Target Validation, and Current Therapeutic Research Programs, have a more complex hierarchy in that Target Validation is further ramified into four subordinate decision criteria, namely Biochemical & Physiological Data, Structural Data, Pharmacological Data, and Medical Data. Importantly, the relative contribution of each of the criteria used to rank the possible targets (*i.e.*, alternatives) with respect to all of the business and scientific criteria, can easily be modified to assess their effect on the relative rankings at the various levels of the hierarchy. This is a type a sensitivity analysis that plays a crucial role in the decision process as will be seen in the sequel. It is also important to stress that this is only one possible view of the relevant business and scientific issues. In fact, the AHP is quite flexible and is well suited to assessing a large number 'what if' scenarios over many different sets of criteria and subcriteria.

First, consider comparative evaluation of the four criteria under Business Issues. Table 3 shows the relative importance attributed to each of the pairs of criteria making up the comparison matrix. The inconsistency index for this matrix is **I**=0.00, that is the comparative ratings of the criteria associated with Business Issues are internally consistent.

Table 3. Comparing Business Criteria.

	Market Potential	Unmet Medical Need	IP Position	External Competition	Priority Ranking
Market Potential	1	3/2	1	2/1	0.269
Unmet Medical Need	2/3	1	1/2	1/2	0.155
IP Position	1	2/1	1	1	0.288
External Competition	1/2	2/1	1	1	0.288

The priority rankings given in the last column of the table are the normalised components of the principal eigenvector, which indicate that IP Position and External Competition are most important followed closely by Market Potential, all three being significantly more important than Unmet Medical Need. As will be seen in the sequel, the comparative values can be easily adjusted and the impact of the adjustments on the overall rankings can be easily assessed. It is important to recognise that the methodology has tremendous flexibility and that both the criteria and their comparative values are subject to modification.

Analogously to Business Issues, the following comparative values make up the comparison matrix for Scientific Issues as shown in Table 4. Unlike for Business Issues, the comparative ratings for the three criteria under Science Issues have an inconsistency index of **I**=0.10, which is near the upper bound of “acceptable” values for this index.

Table 4. Comparing Science Criteria.

	Freedom to Operate	Target Validation	Current Therapeutic Research	Priority Ranking
Freedom to Operate	1	3/2	1/2	0.221
Target Validation	2/3	1	2/1	0.460
Current Therapeutic Research	2/1	1/2	1	0.319

From the table it is clear that Target Validation is the most important criterion followed by Current Therapeutic Research and Freedom to Operate. Target Validation is obviously important as unvalidated targets would be less desirable than validated ones. However, consideration of the nature of the validation is also important. Thus, Target Validation is further ramified in an effort to address the relative importance of the different categories of validation, which will be discussed further below (see also Table 5). Current Therapeutic Research assesses how on-going research projects may impact the choice of new targets. This manifests itself in basically two ways, competition from on-going projects and an improved experience and knowledge base due to research that has been carried out in the area. In contrast to the case of Target Validation these two competing factors will not be explicitly considered, although to do so is quite simple, requiring only an addition level to the hierarchy subsumed under the Current Therapeutic Research category. Freedom to Operate is related to IP Position. IP Position focuses primarily on the patent status of bioactive compounds related to the target and whether there is sufficient room in chemistry space to discover and develop new compounds for the target. Freedom to Operate, on the other hand, focuses more on the patent status of the target itself as well as the related technologies needed to effectively carry out drug discovery research on the target.

A comparison of the criteria relevant to Target Validation are presented in Table 5.

Table 5. Comparing Target Validation Criteria.

	Biochem. & Physiol. Data	Structural Data	Pharmacol. Data	Medical Data	Priority Ranking
Biochem. & Physiol. Data	1	3/1	2/1	3/1	0.463
Structural Data	1/3	1	1	1	0.172
Pharmacol. Data	1/2	1	1	3/2	0.210
Medical Data	1/3	1	2/3	1	0.154

As was the case for Business Issues, the inconsistency index has a value of **I**=0.01, that is the comparative ratings are essentially fully consistent.

From the Priority Ranking column in the table it is clear that Biochemical & Physiological Data is the single most important decision criterion with respect to Target Validation, more than twice as important as any of the other criteria. As noted several times above, the results given in this table represent only one set of comparative judgements. In addition, other criteria may be added or some of the present criteria could be modified or eliminated. These are issues that must be dealt with by the decision makers who possess appropriate domain knowledge.

Global priorities for all of the criteria are given in Table 6. The mathematical expressions for computation of the global priorities are given in the Appendix. Note that these are in general multilinear rather than linear forms. Interestingly, Current Therapeutic Research is significantly more important than any of the other criteria. This is due to the complex chain of weightings from the different levels of the hierarchy, as shown in the Appendix.

Table 6. Global Priorities.

Criterion	Priority Rating
Current Therapeutic Research	0.213
Freedom to Operate	0.147
Biochem. & Physiol. Data	0.142
IP Position	0.096
External Competition	0.096
Market Potential	0.090
Pharmacological Data	0.064
Structural Data	0.053
Unmet Medical Need	0.052
Medical Data	0.047

To determine the overall target rankings it is necessary first to develop a *rating scale* for each of the targets with respect to each of the global priorities. Table 7 illustrates such rating scales for three of the criteria: Unmet Medical Need, External Competition, and Biochemical and Physiological Data. Typically, a rating scale assigns a numerical priority ranking to each object being ranked (targets in the present case) with respect to each of the relevant criteria. A qualitative description is associated with each priority ranking score. For example, under Unmet Medical Need the priority ranking of 1.00 is associated with the descriptive phrase "Very Large," while the score of 0.25 is associated with "Small". The use of such descriptive language to characterise how a given target is ranked with respect to a specific criterion facilitates the type of qualitative reasoning that is essential in many decision making processes and is particularly useful here. Note that the largest priority value is one and that the priority values do not sum to unity. This is called the "Ideal" normalisation and is used in all of the rating scales in this study.

It is important to note that both the description and priority score should be relevant to the criterion being considered.

Table 7. Examples of Rating Scales.

Unmet Medical Need			External Competition			Biochem. & Physiol. Data	
Description	Priority		Description	Priority		Description	Priority
Very Large	1.00		None	1.00		Significant	1.00
Large	0.75		Weak	0.75		Reasonable	0.75
Moderate	0.50		Moderate	0.50		Small	0.50
Small	0.25		Strong	0.25		Very Little	0.25
Very Small	0.01		Very Strong	0.01			

Take for example the case of External Competition where the ordering of descriptions seems to be reversed from the typical ordering seen in Unmet Medical Need or in Biochemical and Physiological Data. In External Competition the description "None" corresponds to a priority value of 1.00, which is quite sensible here since the most desirable case would be one in which external competition is non-existent.

To determine the overall rankings for a specific target, each of the ten global priorities given in Table 6 is multiplied by its corresponding priority score for that target and the products are summed. The results for all nine targets are summarised in Table 8 (see next page), which shows the rank ordering of the targets and the ratings for each criterion. Target #2 is seen to be the highest ranked target and Target #9 the lowest ranked target - note that the ranking is again based upon the Ideal scale. The *relative ratings* (on the unit scale) of the five best targets is given in Table 9. These results have an overall inconsistency index of **I** = 0.02, which is quite good. From the table it is clear that the resulting rankings are reasonably close numerically, which begs the question of exactly how sensitive the final results are to the various choices of the scales and comparative judgements used in the decision model.

EXPERTCHOICE2000™ provides a useful facility for exploring the sensitivity of the decision model to the choice of scales, comparative judgements among criteria, and the presence or absence of specific criteria. Several examples that illustrate the effect of modifying the relative weighting of various criteria on the overall goal of the decision process are provided in Figures 3-5.

Both plots in Figure 3 are concerned with the effect of modifying the four Target Validation variables-Biochemical & Physiological Data, Structural Data, Pharmacological Data, and Medical Data (see Table 5 for the weights and additional details).

Table 8. Rank ordering of the targets and the ratings for each criterion

Ideal Mode		**Ratings**									
Alternatives	Total	Market Potential	Unmet Medical Need	IP Position	External Com-petit.	Freedom to Oper-ate	Bio-chem. & Physiol. Data	Struc-tural Data	Pharma-col. Data	Medical Data	Current Thera-peut. Research Programs
Target #2	0.741	Moder-ate	Moderate	Strong	None	Weak	Reason-able	Very Little	Reason-able	Small	Strong
Target #1	0.734	Large	Small	Moderate	Weak	Moderate	Signifi-cant	Small	Reason-able	Reason-able	Strong
Target #3	0.725	Very Large	Small	Strong	Strong	Moderate	Reason-able	Small	Signifi-cant	Signifi-cant	Strong
Target #4	0.685	Small	Large	Very Strong	Moderate	Weak	Signifi-cant	Reason-able	Small	Small	Modest
Target #5	0.665	Very Small	Very Large	Strong	Very Strong	Strong	Signifi-cant	Signifi-cant	Reason-able	Reason-able	Strong
Target #8	0.582	Moder-ate	Moderate	Very Weak	Moderate	Weak	Reason-able	Small	Reason-able	Small	Modest
Target #7	0.465	Large	Small	Weak	Weak	Weak	Small	Small	Very Little	Very Little	Weak
Target #6	0.431	Moder-ate	Small	Moderate	Weak	Very Strong	Reason-able	Reason-able	Small	Very Little	Weak
Target #9	0.385	Moder-ate	Moderate	Very Strong	Strong	Weak	Small	Very Little	Very Little	Very Little	None

Table 9. Relative ratings of five best targets.

Target Rankings	Relative Ratings
Target #2	0.211
Target #1	0.206
Target #3	0.203
Target #4	0.196
Target #5	0.184

In the upper panel the heights of the unfilled vertical bars correspond to the respective weights used in the current study, namely, 0.463, 0.172, 0.210, and 0.154, for the four variables (see the left-hand ordinate). The colored lines correspond to the normalised ratings of Targets #1–#5 with respect to each of the same four variables (see the right-hand ordinate). Ratings with a large spread of values, such as those associated with Structural Data, tend to be more sensitive than those with a small spread of values, such as Biochemical & Physiological Data, to changes in the relative weightings. The OVERALL values denote the target ratings with respect to Target Validation only.

Comparing the upper and lower panels of the figure clearly shows that by increasing the weight for Structural Data, the most sensitive variable, from 0.17 to 0.48 not only increases the general spread of the ratings values but, more importantly, causes a change in the order of target rankings. Changing the weighting of a less sensitive variable such as Biochemical & Physiological Data has a much smaller overall effect and does not change the ranking order.

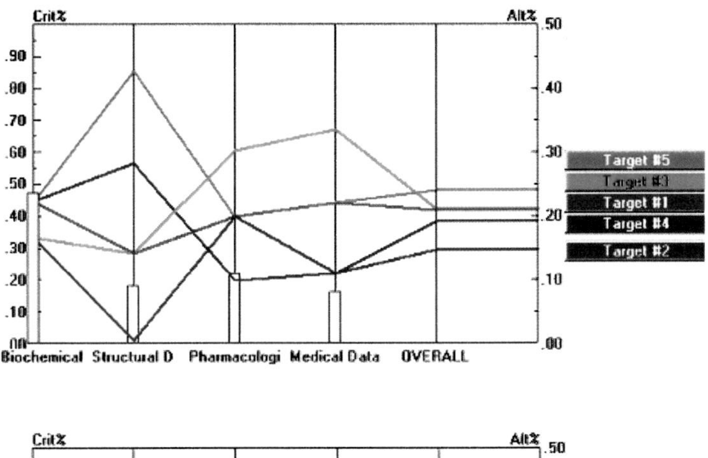

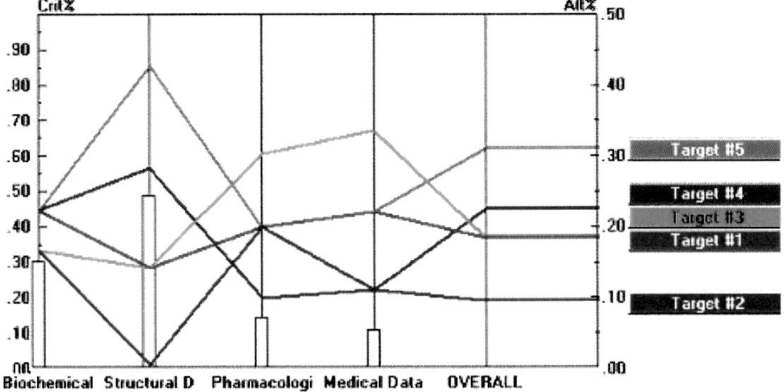

Figure 3. Sensitivity plots from *EXPERTCHOICE2000*™ showing the effect of changing the weighting of Target Validation variables - Biochemical & Physiological Data, Structural Data, Pharmacological Data, and Medical Data - on the overall target rankings. Consult the text for further details.

The upper panel in Figure 4 illustrates the sensitivity of the target rankings to changing the weightings of the four criteria associated with Business Issues - Market Potential, Unmet Medical Need, IP Position, and External Competition (see Tables 3 and 4 for additional details).

As was the case in Figure 3, the coloured lines indicate the relative ratings of the targets with respect to each of the business criteria (upper panel) and science criteria (lower panel), and the unfilled bars indicate their relative weights. Because it has the largest ratings spread, External Competition will have the largest effect on the overall target ratings with respect to Business

Issues, while Freedom to Operate will have a similar, but relatively smaller effect, on the over target ratings with respect to Science Issues.

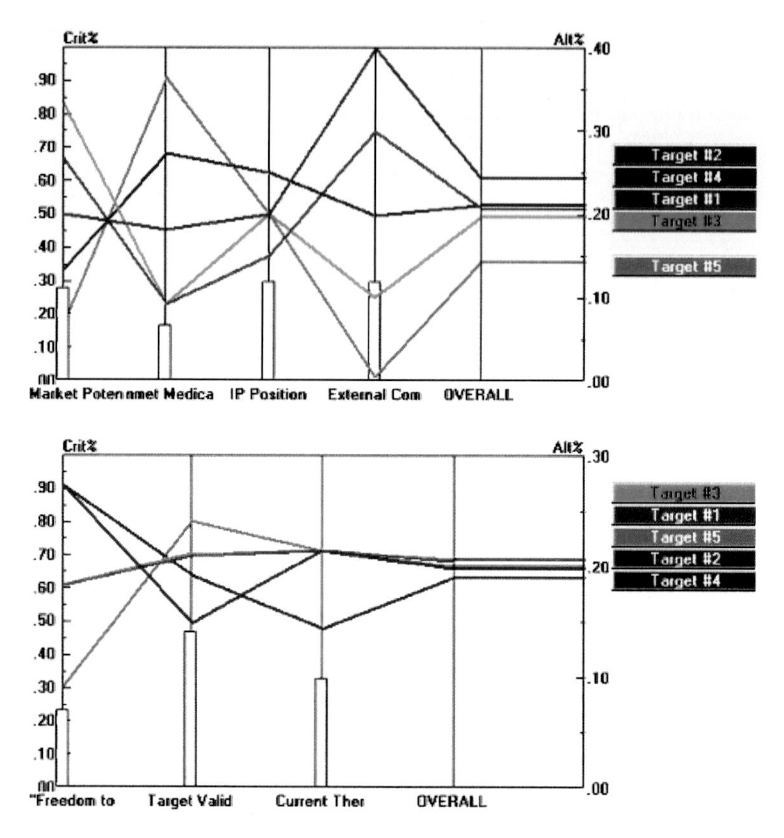

Figure 4. Sensitivity plots from *EXPERTCHOICE2000*™ illustrating the sensitivity of the four business-related issues (upper panel) and the three science-related issues (lower panel). Consult the text for further details

In the upper panel of the Figure 5 the unfilled vertical bars show the original weightings for Business Issues and Science Issues, 0.33 and 0.66, respectively. The coloured lines in the figure correspond to the values that the different targets have with respect to Business Issues and Science Issues. These values are appropriately modified by the weights for Business Issues and Science Issues and then combined to yield the OVERALL scores, which in this case are the final ratings and thus rankings of the different targets. The target rankings shown in colour correspond to the values given in Table 9. The lower panel of the figure shows the effect of modifying the weights so that Business and Science Issues are now of equal importance. As is seen in the figure, target rankings are unchanged although Target #1 is now ranked somewhat higher and Target #5 somewhat lower.

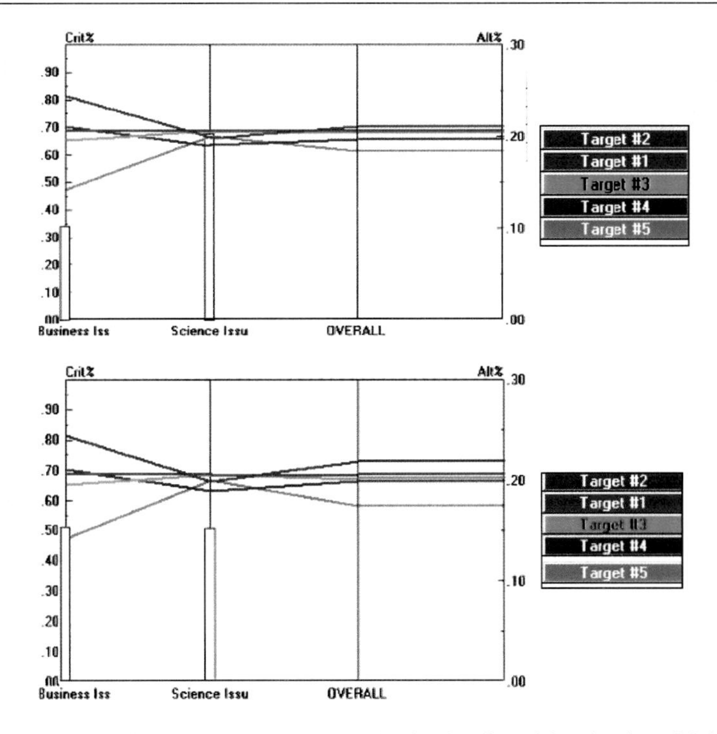

Figure 5. Sensitivity plot from *EXPERTCHOICE2000*™ showing the effect of changing the weighting of Business Issues with respect to Science Issues on the overall target rankings. Consult the text for further details.

The relative rankings of the other targets remain largely unchanged. To change the order of the rankings requires a significant distortion of the Business Issues to Science Issues ratio. Thus, the rankings are largely stable to perturbations of these weightings. This is not, however, the case with respect to other criteria, as was seen above, but this is not surprising given the narrower ratings spreads.

CONCLUSIONS AND FUTURE WORK

As research environments become more and more complex the need for computer-aided decision making methods will gain in importance. As seen in the present work, the AHP is a flexible decision-making tool that is capable of dealing with the types of subjective and objective data that are typically associated with many scientific decisions. Importantly, sensitivity analysis provides an appropriate means for assessing the robustness of a given decision model. It is also important to note that the usefulness and applicability of each decision model depends heavily on the domain knowledge of the decision makers. In fact, the results afforded by any decision model built without appropriate domain knowledge are at best likely to be misleading and at worst likely to be entirely meaningless.

Although the example given above deals only with biological target selection, it contains many of the features found in other scientific decision processes, examples of which include: (1) assessing "molecular quality," (2) evaluating molecular docking software, (3) assessing biological promiscuity, and (4) assessing drug candidate status. An interesting possible application of the AHP methodology may be in assessing the performance of scientific research personnel. While such an application has not to my knowledge been carried out to date, many such assessments have been carried out in a number of business areas.

REFERENCES AND NOTES

[1] Saaty, T. L. (1980). *The Analytic Hierarchy Process.* McGraw-Hill, New York.

[2] Saaty, T. L. (1994). *Fundamentals of Decision Making and Priority Theory.* RWS Publications, Pittsburgh, PA.

[3] Saaty, T. L. & Vargas, L. G. (1994). *Decision Making in Economic, Political, Social, and Technological Environments.* RWS Publications, Pittsburgh, PA.

[4] Saaty, R. W. & Vargas, L. G. (Eds.) (1987). The Analytic Hierarch Process—Theoretical Developments and Some Applications. *Mathematical Modelling* **9**:161-395.

[5] Lootsma, F. A. (1999). *Multi-Criteria Decision Making Via Ratio and Difference Judgement.* Kluwer Academic Publishers, Dordrecht, The Netherlands.

[6] Saaty, T. L. (1996). *The Analytic Network Process—Decision Making with Dependence and Feedback.* RWS Publications, Pittsburgh, PA.

[7] See Table 3.1 in reference (2).

[8] *EXPERTCHOICE2000*, Expert Choice, Inc., Pittsburgh, PA, (2000).

APPENDIX

As noted by Saaty (1-4, 6), the hierarchical, weighted summations carried out in the AHP are not linear forms but are rather more complex mathematical objects called *multilinear forms*. This is illustrated by considering the hierarchy in Figure 2, which is shown again in Figure A1, where all of the explicitly designated decision criteria have been symbolically represented for mathematical convenience. Multilinear forms are constructed from the nested, weighted summations of linear forms such as those given in Eq. (6). This illustrated in Eq. (A1) for A_k (G), the value for the k-th alternative with respect to the overall goal in the hierarchy depicted in Figure A1,

$$\begin{aligned} A_k(G) = & \, w(C_1)\sum_{i=1}^{4} w(C_i^1)\cdot A_k(C_i^1) \\ & w(C_2)\left\{ w(C_1^2)\cdot A_1(C_1^2) + w(C_2^2)\sum_{i=1}^{4} w(C_i^{22})\cdot A_k(C_i^{22}) + w(C_3^2)\cdot A_k(C_3^2)\right\} \end{aligned} \tag{A1}$$

Expanding Eq. (A1) yields

$$\begin{aligned} A_k(G) = & \, w(C_1)\cdot w(C_1^1)\cdot A_k(C_1^1) + w(C_1)\cdot w(C_2^1)\cdot A_k(C_2^1) \\ & + w(C_1)\cdot w(C_3^1)\cdot A_k(C_3^1) + w(C_1)\cdot w(C_4^1)\cdot A_k(C_4^1) \\ & + w(C_2)w(C_1^2)\cdot A_1(C_1^2) \\ & + w(C_2)\cdot w(C_2^2)w(C_1^{22})\cdot A_k(C_1^{22}) + w(C_2)\cdot w(C_2^2)w(C_2^{22})\cdot A_k(C_2^{22}) \\ & + w(C_2)\cdot(C_2)\cdot w(C_2^2)w(C_3^{22})\cdot A_k(C_3^{22}) + w(C_2)\cdot w(C_2^2)w(C_4^{22})\cdot A_k(C_4^{22}) \\ & + w(C_2)w(C_3^2)\cdot A_k(C_3^2) \end{aligned} \tag{A2}$$

The multilinearity comes from the product weight terms terms such as $w(C_2)\cdot w(C_2^2)w(C_1^{22})$. Considering all of the $A_k(G)$ terms, where k=1,2,...,n, yields n equations similar to Eq. (A2), which can be rearranged into the matrix equation shown in Eq. (A3) below

$$\begin{bmatrix} A_1(C_1^1) & A_1(C_2^1) & \cdots & A_1(C_1^{22}) & \cdots & A_1(C_3^2) \\ A_2(C_1^1) & A_2(C_2^1) & \cdots & A_2(C_1^{22}) & \cdots & A_2(C_3^2) \\ \vdots & \vdots & & \vdots & & \vdots \\ A_n(C_1^1) & A_n(C_2^1) & \cdots & A_n(C_1^{22}) & \cdots & A_n(C_3^2) \end{bmatrix} * \begin{bmatrix} w(C_1) * w(C_1^1) \\ w(C_1) * w(C_2^1) \\ \vdots \\ w(C_2) * w(C_2^2) * w(C_1^{22}) \\ \vdots \\ w(C_2) * w(C_3^2) \end{bmatrix} = \begin{bmatrix} A_1(G) \\ A_2(G) \\ \vdots \\ A_n(G) \end{bmatrix} \tag{A3}$$

This is identical in form to Eq. (7) except that the terms in the "weight vector" are multilinear rather than linear.

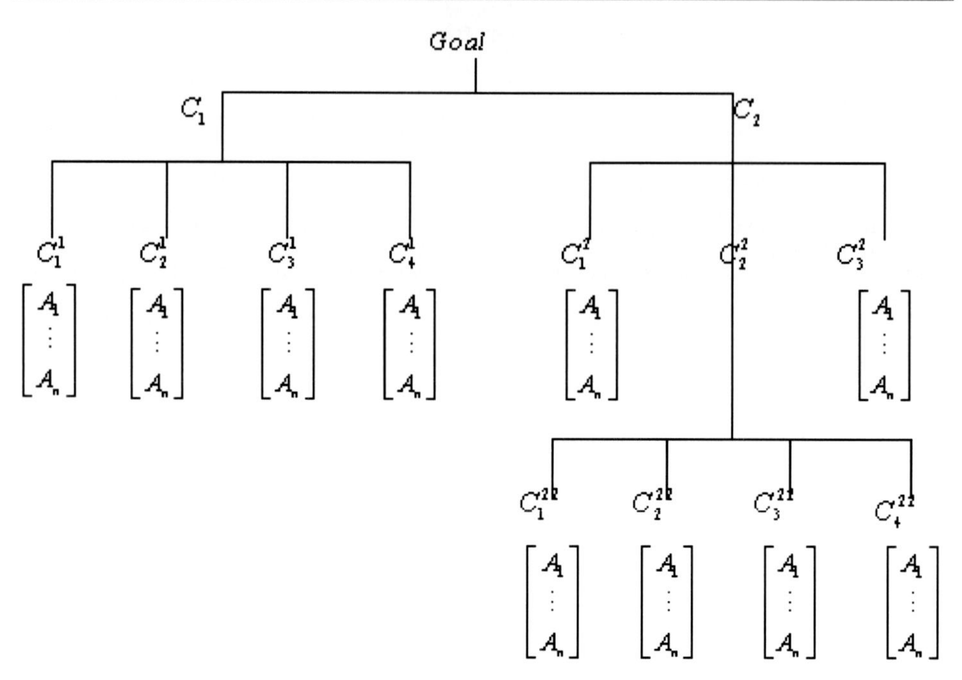

Figure A1. Target-assessment hierarchy identical to Figure 2 except that the designations have been replaced by mathematical symbols.

KNOWLEDGE-BASED LEAD FINDING BY MATCHING CHEMICAL AND BIOLOGICAL SPACE

KARL-HEINZ BARINGHAUS*, THOMAS KLABUNDE, HANS MATTER, THORSTEN NAUMANN AND BERNARD PIRARD

Aventis Pharma Deutschland GmbH, LG Chemistry-Computational Chemistry,
Industriepark Hoechst, Building G 878, D-65926 Frankfurt am Main, Germany
E-Mail: karl-heinz.baringhaus@aventis.com

Received: 3rd July 2002 / Published: 15th May 2003

ABSTRACT

This paper describes a target family-related lead finding approach, which consists of capturing public and proprietary information to build a biological and a chemical space. Computational tools to assemble these spaces as well as appropriate techniques to match them are covered. Three recent applications in the field of kinases, ion channels and GPCRs exhibited already improved lead finding capabilities compared to traditional approaches.

INTRODUCTION

Molecular Informatics is usually involved in several parts of the value chain of drug discovery, mainly of course in lead finding and lead optimization through appropriate techniques, such as for instance HTS data analysis, database mining and pharmacophore modeling (1). However, the competitive pressure in the pharmaceutical industry requires a reduction of project cycle-times and therefore an increase in productivity and efficiency.

In addition, pharmaceutical companies are faced with the challenge of translating genomic information into new, innovative medicines. More than a thousand potential new drug targets have emerged from the sequencing of the human genome, but currently available drugs only target approximately 500 different proteins (2). In order to effectively cope with genomic research, an improvement in lead finding is needed.

A good strategy would be a molecular informatics driven knowledge based approach, whereby chemical information and expertise within target families are acquired and applied. Thereby, a match of chemical and biological information within target families is achieved (3). This article highlights computational methods towards such a target family approach and contains three recent examples in this field.

Matching of Chemical and Biological Space

Our target family related approach consists of biased libraries, which were built by matching chemical and biological space. Therefore, the intersection of biological structures and functionalities with chemical structures and properties is derived to perform a knowledge driven biased library design. This allows an extraction of common structural features for target families out of a more or less infinite chemical space. Thereby, chemical libraries could be built, enriched by preferred features of biologically active compounds (Fig. 1).

Target family related structural features are identified by a combined 2D and 3D analysis.

The 2D approach is based on a collection of biologically active compounds and consists mainly of similarity and substructure searching and of the analysis of common frameworks and fragments.

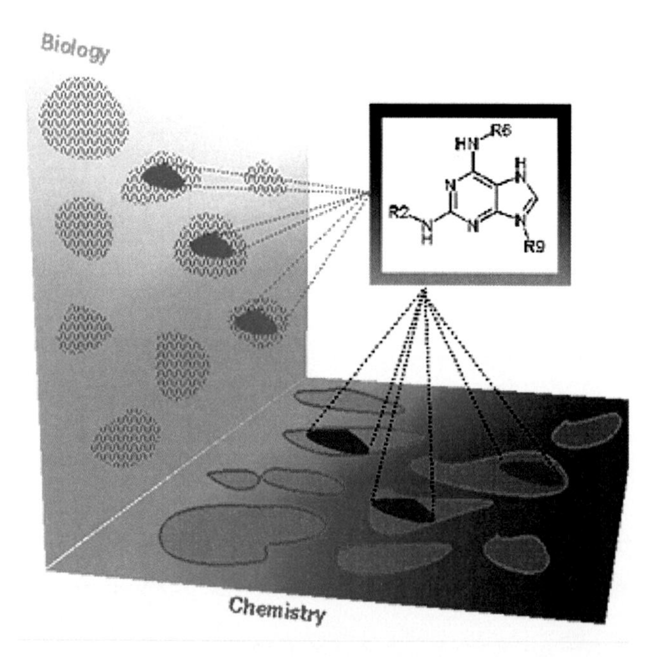

Figure 1. Matching biological structures and functionalities with chemical structures and properties.

The 3D approach relies on homology models, pharmacophores and binding sites, which are used for virtual screening and for our target family landscape approach (Fig. 2).

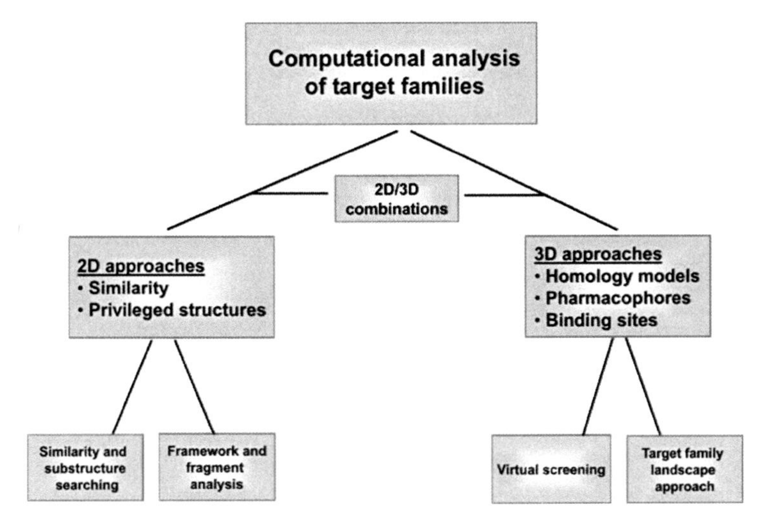

Figure 2. Computational tools to analyze target families for a knowledge driven design approach.

Three of these techniques are outlined here in more detail.

Bemis and Murcko (4) published the topological framework analysis in 1996.

They analyzed shapes of existing drugs from a commercial database to extract drug-related molecular frameworks. A graph theoretical approach was used to decompose molecules into rings and non-cyclic side chains. Linkers and rings together form the framework of a molecule (Fig. 3).

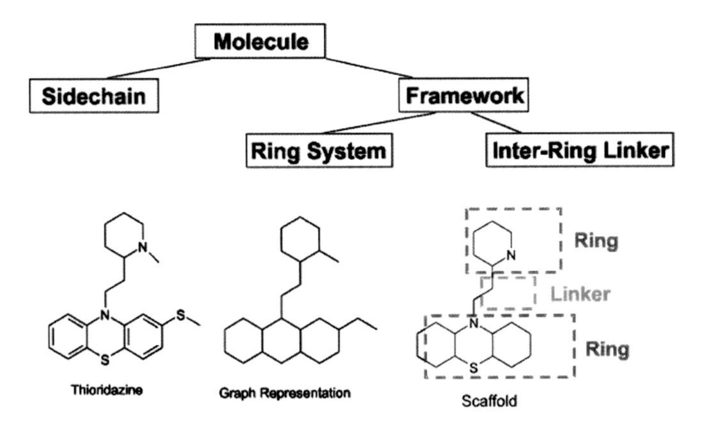

Figure 3. Topological framework analysis.

In this example, acyclic side chains of thioridazine are removed leaving a framework composed of two rings and one interring linker.

Application of such a topological framework analysis on biologically active compounds within a target family reveals access to privileged substructures for activity. By conversion of these frameworks into appropriate scaffolds for synthesis, target family related libraries can be built.

The fragment analysis (5) is based on the RECAP algorithm, published in 1998. This retrosynthetic combinatorial analysis procedure begins with a collection of active molecules and then fragments these molecules using any of the 11 retrosynthetic reactions. For example, Cisapride is cleaved into four fragments based on three different bond cleavage types (Fig. 4).

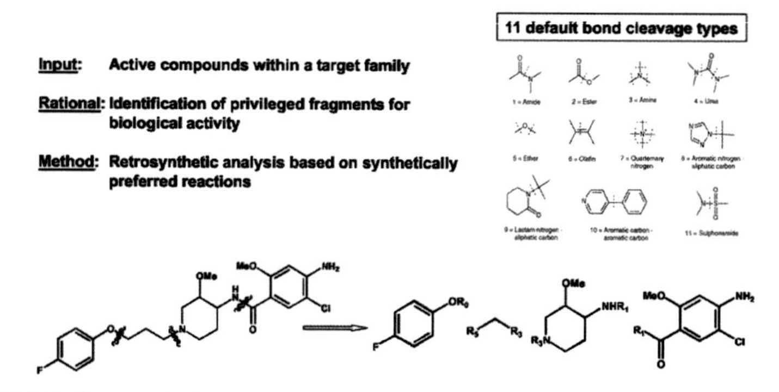

Figure 4. RECAP: Retrosynthetic Combinatorial Analysis Procedure

Resulting fragments are usually clustered and transformed into sets of monomers for subsequent library design. Because the monomers come from biologically active compounds there is a high likelihood that new designed molecules from them will contain biologically interesting motifs.

Our target family landscape approach to classify target family related proteins is a four step procedure (6) (Fig. 5), starting with the selection of representative structures from the Brookhaven and our internal database, followed by an alignment of all structures from 2D sequence and 3D structure similarity. This alignment and all subsequent analysis are focused on the 3D binding site only.

Protein-ligand interactions are then calculated using the GRID force field. They somehow reflect the biological similarity and dissimilarity of proteins from a ligand perspective. A subsequent statistical analysis of the attractive interactions by chemometrical tools, like a principal component analysis, highlights those areas that reveal most differences.

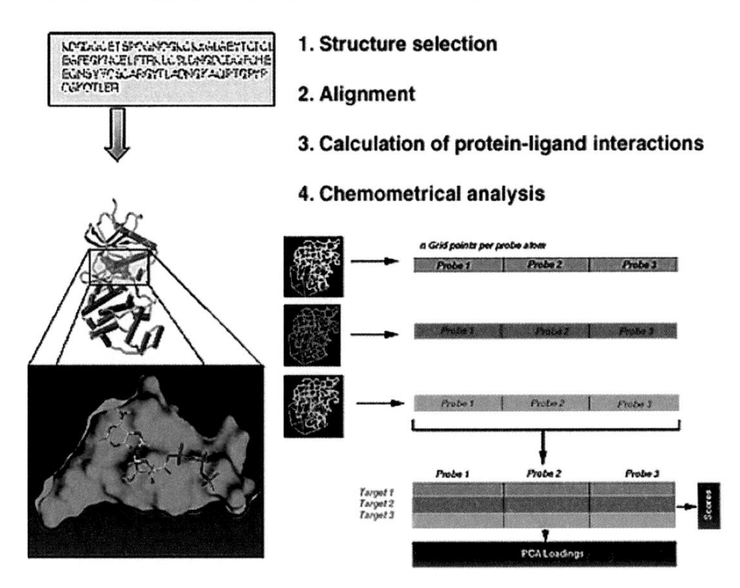

Figure 5. Target family landscape analysis.

This quantitative analysis of GRID-derived molecular interaction fields leads then to a classification of proteins and to a detailed understanding of which structural features are responsible and most characteristic for a particular protein. Hence, this is an excellent description of biological space.

The chemometrical analysis (PCA) of this landscape approach is schematically shown in Figure 5. Starting from the GRID interaction field for one single GRID probe and one protein a vector is constructed. Subsequent interaction fields for other probes are concatenated to this vector to result in a longer vector containing x GRID times n probe points. Similar vectors are derived for all proteins resulting in an X matrix with one single row per protein.

After block scaling to normalize the probe interactions the X matrix is analyzed using PCA or CPCA. Thereby, the data matrix X is approximated by the product of two smaller matrices, scores and loadings. Scores reflect the similarity of proteins, whereas loadings highlight differences in 3D space in terms of molecular interactions (7).

We performed this target family landscape approach to classify kinases based on their ATP binding site (8). We used 26 kinase structures and aligned them to 1ATP, a cyclo AMP dependent kinase as template. GRID interaction fields were calculated through the amide, carbonyl and dry probe. Only attractive interactions were considered for the subsequent chemometrical analysis. The PCA score plot with the first relevant component on the x-axis and the second on the y-axis is shown in Figure 6. We call this plot the target family landscape of kinases.

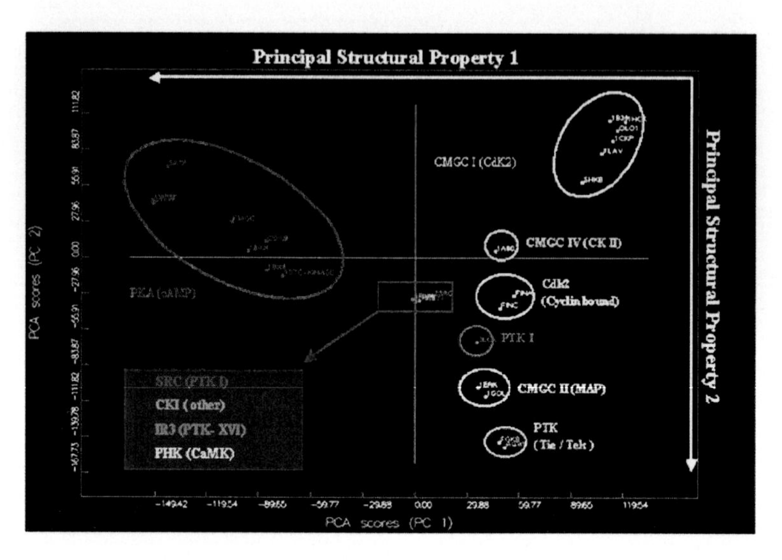

Figure 6. PCA score plot of the kinase landscape.

The first principle component separates CDK and MAP receptor kinases on the right with positive PC1 score values from the family of PKA kinases on the left.

The second principal component allows to separate between MAP and other receptor kinases with negative PC2 scores and the CDK family showing positive PC2 scores.

Kinase subfamily selectivity differences are explained by the corresponding PCA loadings plot (Fig. 7). The loadings of the first principal component of the dry probe clearly describe favorable and selective interactions of the family of PKA and CdK2 kinases. Blue contours indicate hydrophobic regions in space with preferences for PKA, while red contours highlight selective interactions for CdK2 (Fig. 7A). This principle component 1 is dominated by selectivity subsites in the kinase purine and hinge-binding region.

The loadings of the second principal component of the carbonyl probe outline favorable and selective interactions to the CdK2 and PTK1 kinase subfamilies. Blue contours favor hydrogen bond acceptors for CdK2 ligands, whereas red contours highlight selectivity interactions for the PTK1 subfamily (Fig. 7B). This principle component 2 is mainly driven by structural differences in the phosphate binding area.

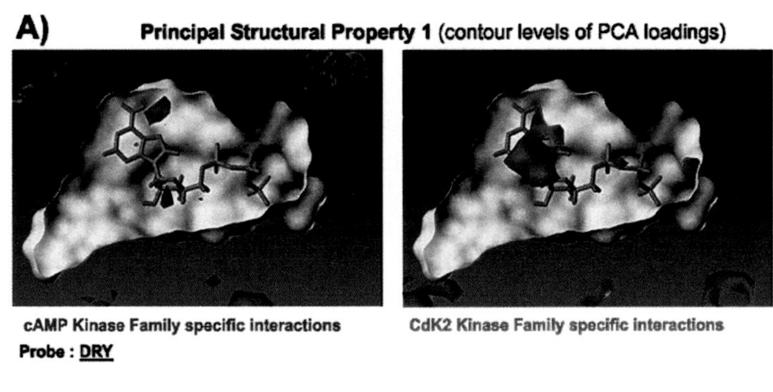

Figure 7. PCA loadings plot of kinases. A) First principle component of the GRID DRY probe highlights selective interactions for PKA and CDK2 kinases. B) Second principle component of the Carbonyl probe differentiates between CdK2 and PTK1 kinases.

EXAMPLES: KINASES, ION CHANNELS AND GPCRS

The design of our kinase library is based on the derived kinase landscape, on pharmacophore models and on privileged frameworks and fragments, which were derived from proprietary and public kinase projects. This kinase specific information was then applied to build our first kinase directed library by cherry picking in our compound collection and by purchasing additional samples (Fig. 8). We are currently and continuously improving this biased library by designing new proprietary kinase scaffolds, from which small compound libraries are built.

Screening of this focused library against new kinase targets like NIK, yielded almost 10-times higher hit rates compared to whole library screening. Such focused screening enables derivation of initial SAR models suitable for subsequent compound optimization. Virtual screening then gives access to the entire compound collection.

Such derived kinase knowledge is applicable in lead optimization of compounds as well. For instance, our initial IκB lead compound was lacking kinase specificity.

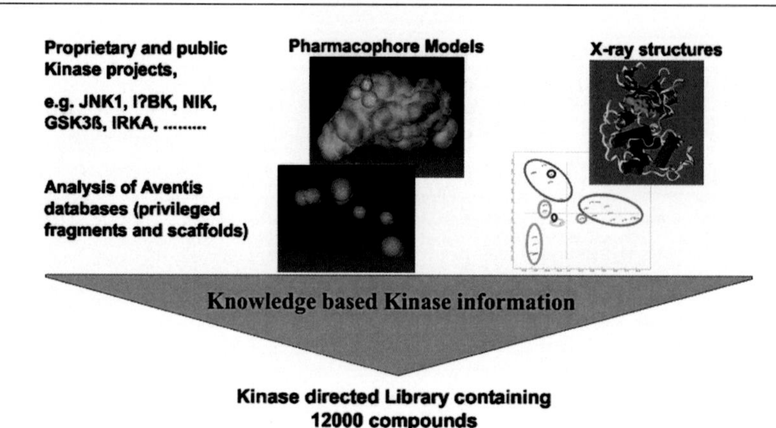

Figure 8. Design of a kinase focused library.

We were able to improve this selectivity by applying our kinase landscape. In this way, we identified a hydrophobic pocket, which is most important to gain selectivity.

Ion channels are of potential interest not only for cardiovascular diseases. However, appropriate high throughput assays to test several hundred thousand compounds against a particular ion channel are still lacking sufficient signal to noise ratios (9). Therefore, a biased ion channel library is of high interest for lead finding in that field.

Our ion channel library, containing 15412 compounds, was composed on public and proprietary ion channel lead compounds. Similarity searches and virtual screening based on pharmacophores and homology models yielded most representatives of this library. Addition of several small combinatorial libraries of ion channel privileged scaffolds then leads to our biased library (Fig. 9). Such QSAR driven refined models or related pharmacophores are well suited for a knowledge-based optimization of certain potassium channel inhibitors. We are constantly improving these models and adding new models as well.

G-protein coupled receptors comprise a large protein family sharing a conserved trans-membrane structure composed of seven trans-membrane helices (11). GPCRs are located at the surface of the cell and are responsible for the transduction of an endogenous signal into an intracellular response. The natural ligands of this receptor family are extremely diverse, for instance biogenic amines, amino acids, lipids, peptides and proteins or nucleosides and nucleotides.

As GPCRs are quite important biological targets, we decided to improve our lead finding capabilities in this field by building a GPCR biased library.

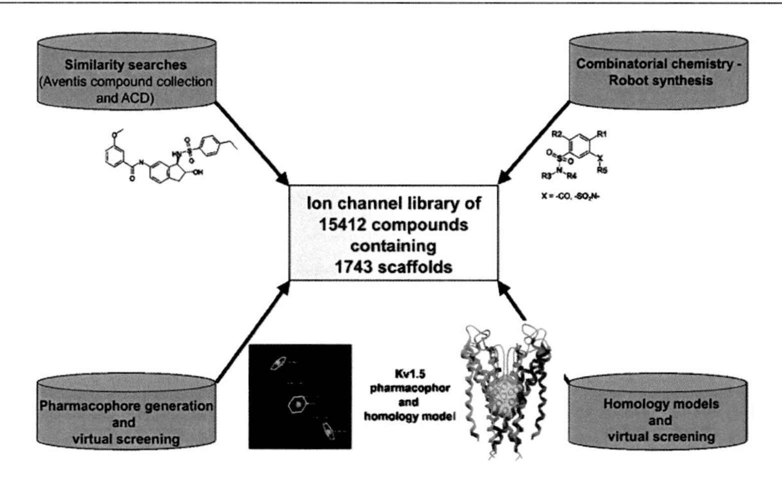

Figure 9. Building of the ion channel library.

Our knowledge based Aventis GPCR library was built upon a GPCR database containing 20,000 public and proprietary GPCR ligands. Fragment and framework analysis revealed privileged GPCR motifs from which some were turned into small combinatorial libaries. Further compound selection by virtual screening yielded our GPCR biased library (Fig. 10). This library still needs improvement, because of the high diversity of potential GPCR ligands.

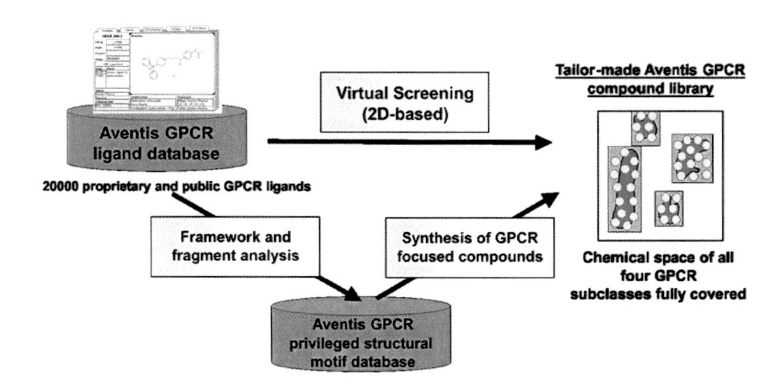

Figure 10. 2D-Based Approach to a Targeted GPCR Library.

The overall outcome of our lead finding strategy of Urotensin II receptor antagonists is outlined in Table 1. We observed an almost 20-times higher success rate by our rational approach in comparison to HTS. The hit rate of our GPCR directed library, however, was only slightly higher than by random screening. This is not disappointing, because that initial GPCR biased library was lacking ligands against peptidic GPCRs.

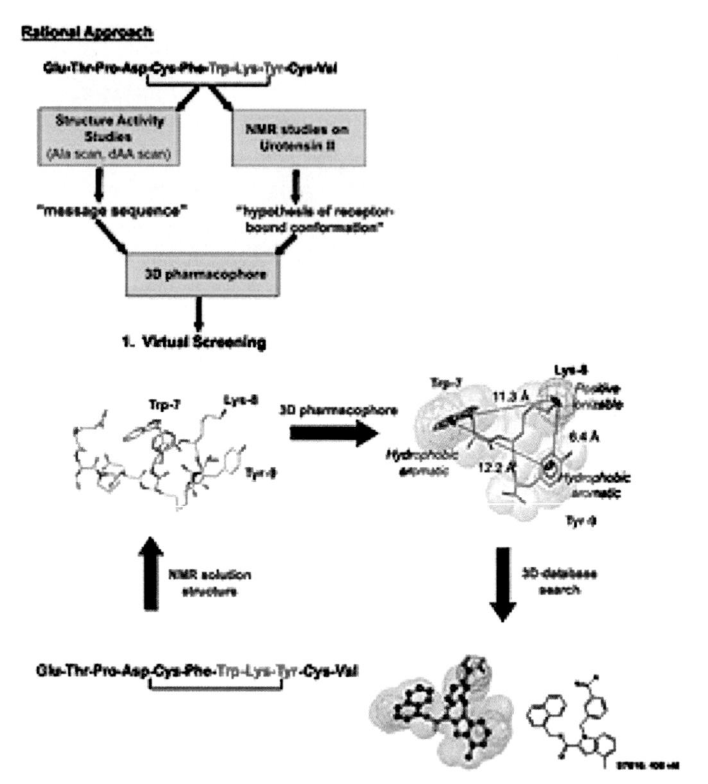

Figure 11. Rational Approach to Identification of Non-Peptidic Urotensin II Receptor Antagonists

We applied recently a knowledge-based lead finding strategy in the GPCR field. We were particularly interested to identify non-peptidic Urotensin II receptor antagonists. Urotensin II, a peptide comprised of 11 amino acids and a disulfide bridge, is the most potent vasoconstrictor known and therefore is of therapeutic interest (12). A combined rational and screening lead finding strategy was applied, whereby a part of our whole library and of course our GPCR biased library was screened. In addition, a rational approach by gathering structure-activity information through an alanine scan and by NMR studies on Urotensin II was performed (Fig. 11). Thereby, the message sequence tryptophan, lysine and tyrosine was successfully identified and a hypothetical receptor bound conformation of Urotensin II was elucidated, from which a 3D pharmacophore was constructed for virtual screening (13).

The spatial 3-dimensional arrangement of the message sequence tryptophan, lysine and tyrosine was elucidated from the NMR solution structure of Urotensin II.

By assuming this as the bioactive conformation a 3-point pharmacophore was built, containing two hydrophobic features from the aromatic moieties of tryptophan and tyrosine and a positive ionizable center from the terminal amino group of lysine. In addition to these three features, the shape of the message sequence was included in the pharmacophore. Subsequent virtual screening returned S7616 as most active Urotensin II receptor antagonist with an IC_{50} of 400nM (Fig. 11).

CONCLUSIONS

Knowledge based lead finding is achieved by matching chemical and biological space through collecting target family related ligands and identification of privileged structural motifs, building distinct 3D pharmacophores, and 3D classification of binding sites to build target family biased libraries for focused screening to find better hits faster.

ACKNOWLEDGEMENTS

The authors thank Clemens Giegerich, Anna Gorokhov, Sven Grüneberg, Gerhard Heßler, Robert Jäger, Andreas Kugelstadt and Stefania Pfeiffer-Marek for fruitful discussions. We also like to thank our collaborators, Gabriele Cruciani from Perugia, Gerhard Klebe and the Cambridge Crystallographic data center, the GMD, now Fraunhofer institute, and BioSolveIT.

REFERENCES AND NOTES

[1] Bajorath, J. (2001). Rational drug discovery revisited: interfacing experimental programs with bio and chemo-informatics. *Drug Discov. Today* **6**:989-995.

[2] Drews, J. & Ryser, S. (1997). The role of innovation in drug development. *Nat. Biotechnol.* **15**:1318-1319.

[3] Wess, G., Urmann, M., Sickenberger, B. (2001). Medizinische Chemie: Herausforderungen und Chancen. *Angew. Chem.* **113**:3443-3453.

[4] Murcko, M. A. & Bemis, G. A. (1996). The properties of known drugs: 1. Molecular frameworks. *J. Med. Chem.* **39**:2887-2893.

[5] Lewell, X. Q., Judd, D. B., Watson, S. P., Hann, M .M. (1998). RECAP-Retrosynthetic combinatorial analysis procedure: A powerful new technique for identifying privileged molecular fragments with usefule applications in combinatorial chemistry. *J. Chem. Inf. Comput. Sci.* **38**:511-522.

[6] Matter, H. & Schwab, W. (2000). A view on affinity and selectivity of nonpeptidic matrix metalloproteinase inhibitors from the perspective of ligands and target. In *Molecular Modeling and Prediction of Bioactivity*; Gundertofte, K., Jorgensen, F. S., Eds.; Kluwer: New York, pp 123-128.

[7] Westerhuis, J. A., Kourti, T., Macgregor, J. F. (1998). Analysis of multiblock and hierarchical PCA and PLS models. *J. Chemom.* **12**:301-321.

[8] Kastenholz, M. A., Pastor, M., Cruciani, G., Haaksma, E. E. J., Fox, T. (2000). GRID/CPCA: A new computational tool to design slective ligands. *J. Med. Chem.* **43**:3033-3044.

[9] Naumann, T. & Matter, H. (2002). Structural classification of protein kinases using 3D molecular interaction field analysis of their ligand binding sites: Target family landscapes. *J. Med. Chem.* **45**:2366-2378.

[10] Xu, J., Wang, X., Ensign, B., Li, M., Wu, L., Guia, A., Xu, J. (2001). Ion-channel assay technologies: quo vadis? *Drug Discov. Today* **6**:1278-1287.

[11] Doyle, D. A., Cabral, J. M., Pfuetzner, R. A., Kuo, A., Gulbis, J. M., Cohen, S. L., Chait, B. T., MacKinnon, R. (1998). The structure of the potassium channel: Molecular basis of K^+ conduction and selectivity. *Science* **280**:69-77.

[12] Flower, D. R. (1999). Modelling G-protein-coupled receptors for drug design. *Biochimica et Biophysica Acta* **1422**:207-234.

[13] Wess, J. (1998). Molecular Basis of Receptor/G-Protein-Coupling Selectivity. *J. Pharmacol. Ther.* **80**:231-264.

[14] Ames, R. S., Sarau, H. M., Chambers, J. K., Willette, R. N., Aiyar, N. V., Romanic, A. M., Louden, C. S., Foley, J. J., Sauermelch, C. F., Coatney, R. W., Ao, Z., Glover, G. J., Wilson, S., McNulty, D. E., Ellis, C. E., Elshourbagy, N. A., Shabon, U., Trill, J. J., Hay, D. W. P., Ohlstein, E. H., Bergsma, D. J., Douglas, S. A. (1999). Human Urotensin-II is a potent vasoconstrictor and agonist for the orphan receptor GPR14. *Nature* **401**:282-286.

[15] Flohr, S., Kurz, M., Kostenis, E., Brkovich, A., Fournier, A., Klabunde, T. (2002). Identification of nonpeptide Urotensin II receptor antagonists by virtual screening based on a pharmacophore model derived from structure-activity relationships and nuclear magnetic resonance studies on Urotensin II. *J. Med. Chem.* **45**:1799-1805.

PROVIDING CHEMINFORMATICS SOLUTIONS TO SUPPORT DRUG DISCOVERY DECISIONS

CARLETON R. SAGE, KEVIN R. HOLME, NIANISH SUD AND RUDY POTENZONE

Lionbioscience, Inc., 9880 Campus Point Dr., San Diego, CA 92121, USA

E-Mail: rudolph.potenzone@lionbioscience.com

Received: 1st August 2002 /Published: 15th May 2003

ABSTRACT

Drug discovery programs have had to deal with an avalanche of data coming from both the adoption of new technologies such as high throughput screening and combinatorial chemistry, as well as advances in genomics and structural genomics which have facilitated a gene family target approach to drug discovery. Although this data rich environment has been a challenge to manage, it has provided an opportunity for the development of informatics based tools and solutions to extract information from this large body of data, and convert this information into knowledge that can be used and reused for drug discovery.

In the cheminformatics field there has been considerable focus on the development of new tools to visualise and analyse the data, particularly with relation to identifying new leads, and analysing SAR for lead optimisation. While individual cheminformatics tools are critical for analysing this data, a real opportunity exists to provide solutions that synthesize results from these analyses into knowledge to support drug discovery decisions. This remains largely a "manual" activity that takes place within individual project teams.

This paper will describe some concepts and implementations of cheminformatics solutions that begin to address the need for reusable knowledge generation within drug discovery projects. The talk will address requirements for the integration of chemical and biological data as well as the integration of tools and models. The power of using predictive tools for compound design will be highlighted as well as methods to simultaneously consider multiple SAR's. We will describe how providing such solutions

INTRODUCTION

The drug discovery process used to be less complicated. Teams of chemists and biologists (drug discovery scientists) would work together on tens to hundreds of molecules to try to specifically alter the function of their biological target. Things have changed. Because of changes in available technologies and increases in fundamental understandings of biology, the drug discovery scientist has to contend with thousands to millions of molecules interacting with potentially thousands of targets. Therefore, the modern drug discovery scientist is awash in data. However, the changes in available technologies haven't necessarily resulted in improvements in the quality of the data. As a result, we have been flung into what might be a morass of meaningless data, or discovery knowledge nirvana. How do we navigate?

COMPONENTS REQUIRED TO BUILD THE MAP

The first stage to approaching this information overload is by assembling the informatics components necessary for integrating all the available data. Databases are an integral component. This simple component may pose problems for some organisations since the default drug discovery data repository has become Microsoft Excel. Once the data has been arranged into databases, establishing a link between the data in the databases is an essential component. This component does not have a trivial solution, since it involves linking data of different types across different research areas. One obvious potential solution is to use the experimental assay data as the common link between the genomics/bioinformatics/proteomics data and cheminformatics data. Assuming that the data has been integrated to some extent, the next required components are ways to visualise, analyse, and interrogate the data. Furthermore, sophisticated computational approaches must also be taken in order to summarise collections of data into reusable knowledge for prospective application. Finally, means to allow `rollup' of these components for decision support, including tracking and knowledge management tools are essential for efficient. Figure 1 illustrates the Lion Bioscience architecture.

Once an integrated system of data, analysis, and decision support tools has been created, then the true power of the system can be exploited for more rational/informed decision making. These systems can be used in all phases of the drug discovery process from target selection to screening set selection to lead optimisation and lead selection for pre-clinical development.

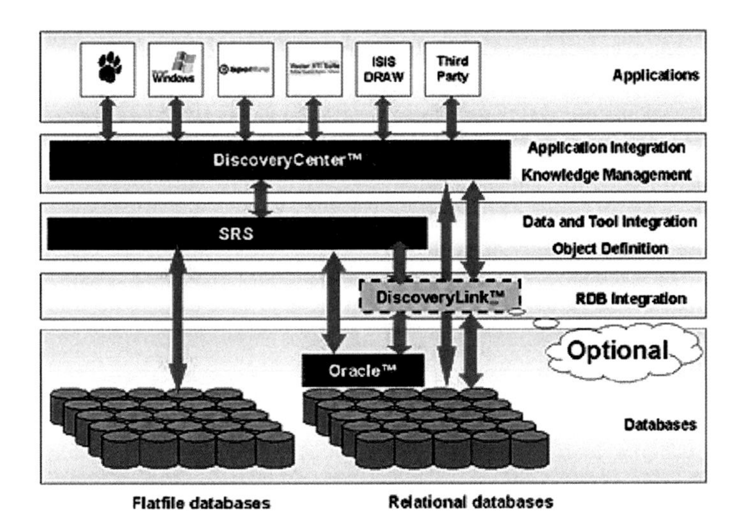

Figure 1. Lion Bioscience integration architecture

The purpose of this paper is to illustrate the potential utility of these approaches in the post-target selection region of drug discovery, and the next two figures illustrate a framework for discussing the small molecule drug discovery process. Figure 2 defines our use of the terms "Lead Identification" and "Lead Optimisation" since these terms may have different connotations in different organisations.

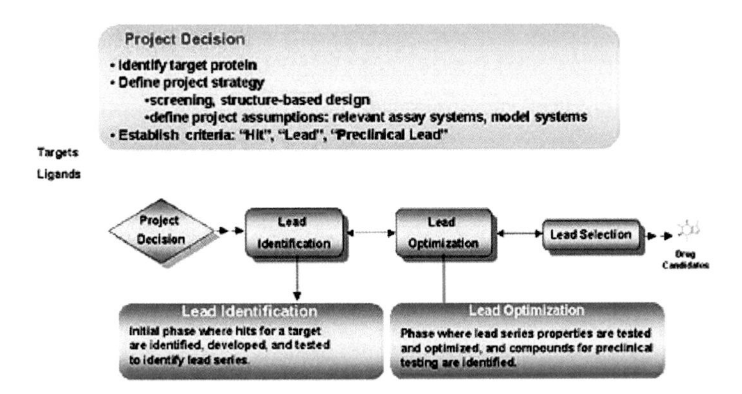

Figure 2. Definitions and Assumptions for Drug Discovery project initiation.

A second concept illustrated in figure 2 are some of the operating assumptions and project criteria that must be performed in order for this integrated approach to work most efficiently. Project criteria are a key decision support / project tracking consideration. Figure 3 illustrates a version of a pre-clinical drug discovery (or lead candidate selection) workflow. It is important to note that this entire workflow sits upon the foundation of knowledge/information gleaned from other projects, and any and all data (both positive and negative) and knowledge generated for a given project in this workflow is put back into this foundational "Knowledge Base".

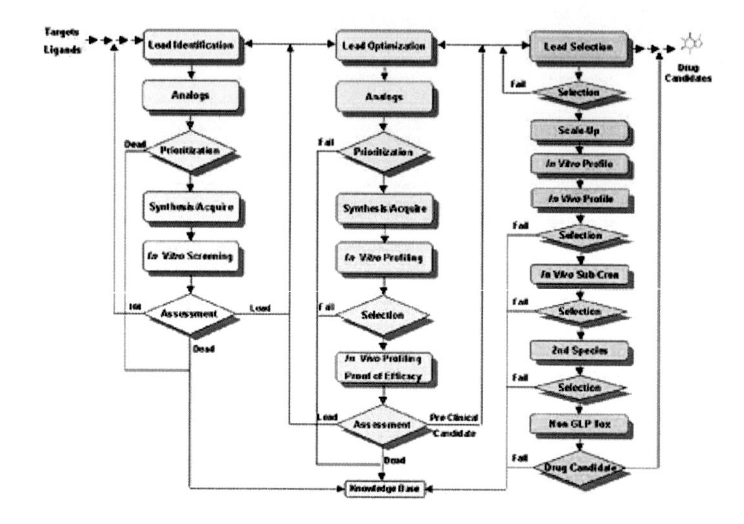

Figure 3. Idealised project work/information flow for Pre-clinical Drug Discovery.

REUSING AND LEVERAGING EXISTING DATA THROUGH COMPUTATIONAL MODELS (HYPOTHESES)

Leveraging data that exists at project initiation via computational means provides momentum in the rational decision making early in the drug discovery process, and continues this momentum as more data becomes available. The type, quantity, and quality of these data define the extent to which computational approaches should be used. Figure 4 describes the potential computational approaches that could be used in support of drug discovery projects given defined sets of experimental data.

In the extreme, if the project under consideration has a liganded protein structure, related protein structures and lead series data available, computational approaches including diversity, similarity, QSAR/pharmacophore models, drug-likeness models, docking, cross-target (or specificity) alert models, and project-specific ADME-models can be applied simultaneously to best leverage all existing information in the compound prioritisation and compound assessment phases of lead identification and lead optimisation (Figure 3).

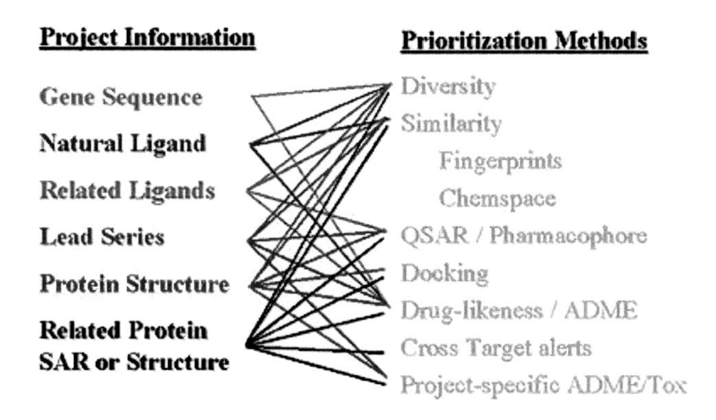

Figure 4. An example of an "available data / available application" matrix

ACCESSING INTEGRATED DATA

Once an integrated data/query system has been created, it must be delivered to the end user in such a way that it actually enhances their work. Here the major challenge is a "people issue". The interface must be simple and familiar, and should not require too much specific additional training for the user to start using it. In our solution to this problem, we have chosen web pages as an interface since it is familiar. has simple interfaces, and is easy to access. Figure 5 represents a "front door" to the system, the place where the end user chooses what task they want to accomplish. Figure 6 demonstrates a simple interface for searching multiple databases simultaneously using an sd, molfile, or sketched molecule.

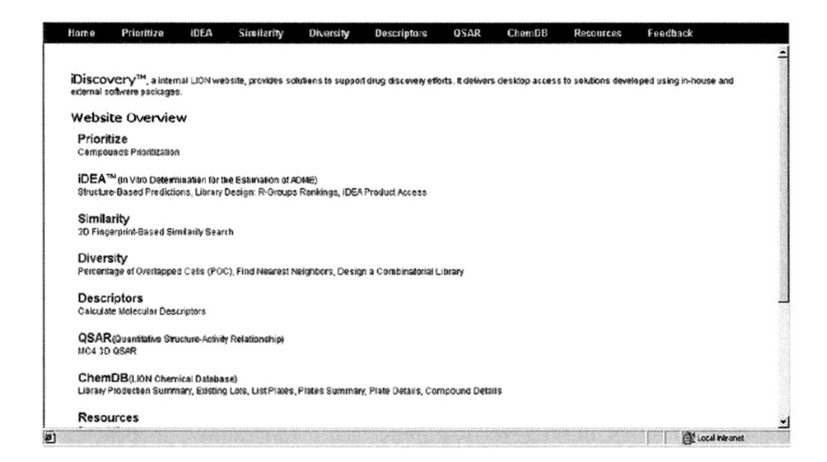

Figure 5. Simple "front page" access to a set of integrated data, models, and tools.

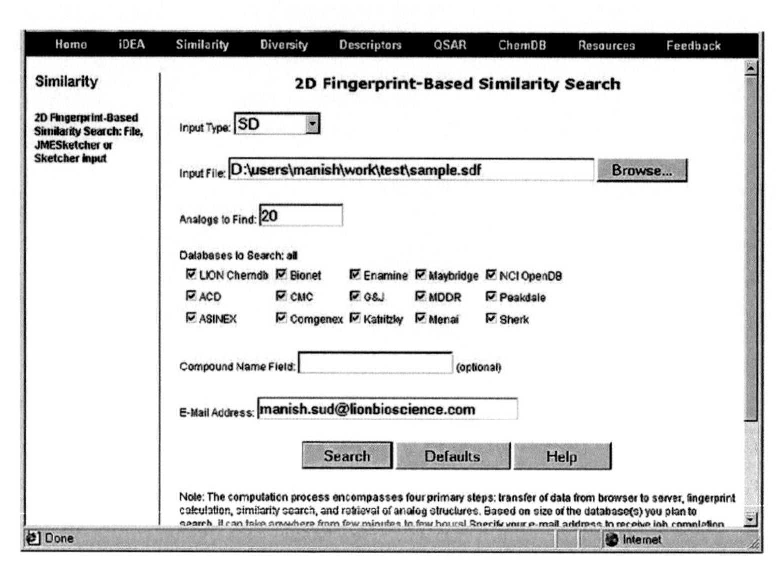

Figure 6. Simple interface for performing a 2D similarity search of multiple databases simultaneously.

TURNING DATA INTO SHARABLE KNOWLEDGE USING COMPUTATIONAL TOOLS

Figure 4 shows the computational methods available given a starting set of experimental data. These approaches have been powerful in enabling complicated hypothesis-driven experimental design in drug discovery project. However, the resultant models are usually "put on the shelf' as projects are promoted or discontinued, leaving these synthesised data unused for future projects. Having these computational models available, and using them appropriately could prove very valuable in addressing specificity, ADME, and safety information of current and future projects, especially in the case of organisations pursuing the target-class approach to drug discovery. Figure 7 shows a matrix of potency/specificity receptor-relevant chemspace (Pearlman reference) models created for a collection of nuclear receptors, representing a collection of easily applicable knowledge about a large fraction of proteins in the nuclear receptor target family. Though we have presented receptor-relevant chemspace models as our in silico surrogates for potency/selectivity, any computationally derived model, from similarity clustering methods to docking may be assembled, integrated, and applied in a cross-project manner.

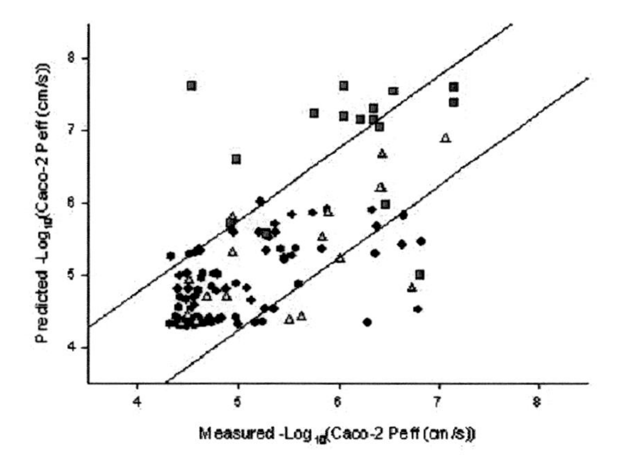

Legend

- ● 0 members outside box
- △ 1-3 members outside box
- ■ >4 members outside box

Figure 7. Distance matrix representing the relatedness between receptor relevant chemspace models.

APPROPRIATELY USING DATA IN THE FORM OF COMPUTATIONAL MODELS

Having data available in an integrated, searchable, analysable context, should be very valuable, however, more value could be added to this data by the creation of robust computational summaries (models) of the data, and applying them in appropriate ways. As an example, may varied approaches and algorithms exist that take a set of compounds and experimental activity data and derive a QSAR model that can often accurately predict the potential activity of a new compound.

It should be possible to reuse these models as a component of a knowledge environment for in silico evaluation of every compound against a surrogate for every potential experimental assay. However, most of the approaches used to build these computational models are statistical in nature, and therefore the performance of these models is only interpolative in nature, therefore, the models will likely perform poorly outside of "chemical space" (extrapolation) they were built upon. An approach to addressing this problem is illustrated in Figure 8 which shows the results from a model to predict Caco-2 effective permeability, including measures of uncertainty, using only the chemical structure of the compound as the only input. This caco-2 model was built using sophisticated statistical pattern recognition methods in which the output is a consensus prediction from 10 independent models. Each model is trained on a different representation of chemical space. To calculate the prospective measure of uncertainty (M.O.U) for the model, the upper and lower bounds of the chemical features (chemical descriptors) were determined to represent the bounds of the multidimensional chemical space upon which each model was trained. For each of the 10 independent (child) models, a new compound may have features whose values are outside of he bounds of the training set.

Nuclear Receptor Average Nearest Neighbor Distance Matrix

Nuclear Receptor Relevant Chemspaces (columns); Nuclear Receptor Probe Ligands (rows)

	ERRa'	ERa	FXR	GCR'	LXR	PPARa'	PPARd	PPARg	PXR'	RARa	RARb	RARg	RORa'	RXRa	RXRb	RXRg	TR'	VDR
ERRa	0.00	0.53	1.05	4.11	1.66	3.33	4.46	4.39	3.47	2.43	2.42	1.84	5.98	2.53	2.53	2.53	3.02	1.21
ERa	1.97	0.00	0.66	4.34	1.78	1.46	2.37	1.65	4.61	1.47	1.41	1.54	6.39	1.09	1.09	1.09	2.07	1.92
FXR	3.93	0.34	0.00	4.12	2.31	1.34	1.78	1.54	5.17	2.27	2.00	2.67	5.82	2.55	2.56	2.56	1.61	3.15
GCR	5.87	1.67	2.17	0.00	1.25	3.19	3.65	3.12	2.97	2.72	2.72	2.08	3.43	1.91	1.91	2.29	2.54	0.57
LXR	3.67	0.87	2.27	1.61	0.00	4.14	5.17	4.45	1.46	3.44	3.43	3.06	3.62	3.60	3.61	3.51	2.12	0.66
PPARa	3.51	0.92	1.06	4.42	3.22	0.00	1.60	1.02	4.95	2.31	2.10	2.24	5.70	2.13	2.17	2.17	1.77	2.83
PPARd	3.53	0.67	0.68	4.66	3.28	0.47	0.00	0.73	5.12	2.06	1.88	2.42	5.92	2.22	2.28	2.28	1.37	3.25
PPARg	4.34	1.64	1.81	4.19	3.67	1.26	2.98	0.00	4.48	3.05	2.85	2.37	4.70	2.00	2.00	2.00	2.21	2.68
PXR	4.71	2.22	3.05	2.26	1.44	4.33	5.26	4.20	0.00	3.81	3.81	2.96	3.74	3.44	3.44	3.64	1.72	1.12
RARa	4.14	0.77	0.50	2.92	1.10	1.37	2.27	1.90	3.78	0.00	0.05	0.16	5.70	0.30	0.36	0.36	3.06	1.23
RARb	4.33	0.69	0.43	2.93	0.97	1.51	2.37	1.80	3.76	0.19	0.00	0.20	5.74	0.23	0.30	0.30	3.26	1.17
RARg	4.14	0.68	0.70	2.85	1.10	1.48	2.43	1.93	3.88	0.30	0.28	0.00	5.66	0.24	0.33	0.34	3.16	1.20
RORa	5.04	2.49	4.45	3.33	3.13	2.60	5.51	2.25	3.81	5.45	5.37	3.04	0.00	1.95	1.95	1.96	3.26	2.52
RXRa	3.99	0.54	1.00	2.90	0.98	1.77	3.06	2.13	3.72	0.87	0.87	0.21	5.13	0.00	0.04	0.15	3.24	1.29
RXRb	4.05	0.71	0.95	2.87	1.19	1.74	2.84	2.03	3.65	0.76	0.74	0.20	5.13	0.02	0.00	0.10	3.25	1.26
RXRg	3.70	0.62	0.76	3.04	1.17	1.66	2.74	2.03	3.69	0.49	0.48	0.12	5.27	0.02	0.00	0.00	3.27	1.23
TR	3.21	1.37	2.12	2.60	1.76	3.03	4.24	3.54	2.63	3.17	3.00	3.17	4.39	3.21	3.22	3.27	0.00	0.90
VDR	2.89	1.27	2.42	1.85	0.88	4.01	5.06	4.55	1.69	3.39	3.38	3.33	3.90	3.67	3.67	3.68	0.88	0.00

Figure 8. Example application of prospective measures of uncertainty in a predictive model for caco-2 effective permeability.

If the features are outside of the bounds for any feature for a child model, that model is considered to be extrapolating. Figure 7 shows an example of the consequences of using extrapolated increases, the error in prediction increases dramatically. Developing methods to assess prediction confidence for all models could aid dramatically both in their successful first-time use as well as enable appropriate reuse of the knowledge gleaned by their creation.

Simultaneous Use of Computational Models and Integrated Data for Compound Prioritisation and Assessment

During Lead Identification and Lead Optimisation, prioritisation or ranking of compounds to acquire, plate, or synthesise can be cumbersome, and is often performed without using all available information. Similarly, assessment of which compound or compound series to promote to the next phase of research should also be performed using all available information simultaneously. Figure 9 shows a simple compound prioritisation input screen for simultaneous prediction of ADME and potency/specificity properties. In this input screen the user may choose which properties to calculate, the criteria defining whether a given compound passes or fails, and the weight of that property in the calculation of a summary score for a molecule.

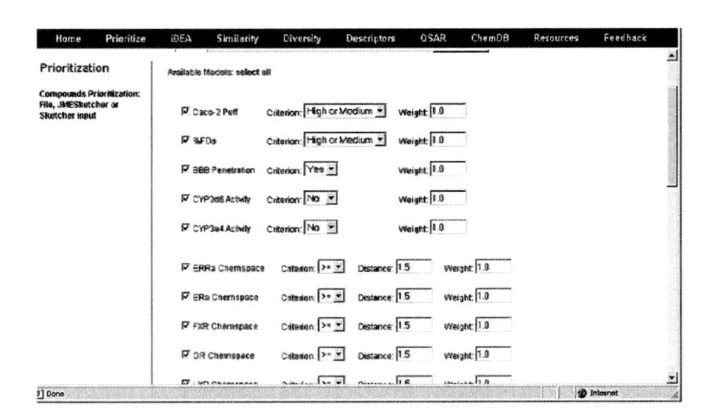

Figure 9. Prioritisation Input Screen. The end user selects the models to run, the criteria required to pass, and the weight each model contributes to the final score.

The summary score allows the composite ranking of all compound under consideration using the same objective evaluation criterion. Figure 10 shows the results of a prioritisation in two views. First, a compound-by-compound view, and second, a property distribution view, which would likely be useful for selecting from commercially available compound collections of in the evaluation of whether or not to synthesise a particular series of compounds versus another. By including integrated data determined experimentally in this analysis, compoung assessment can also be performed.

Application of Integrated Approaches as a Cheminformatics Solution for Decision Support in Drug Discovery: A Hypothetical Estrogen Receptor Project

For this hypothetical example, we will assume that we have chosen Estrogen Receptor alpha (Era) as the target of our Lead Identification (LI)/Lead Optimisation (LO) project. We perform a search of the literature, and find several examples of compounds which have been shown to interact with Era (Figure 11).

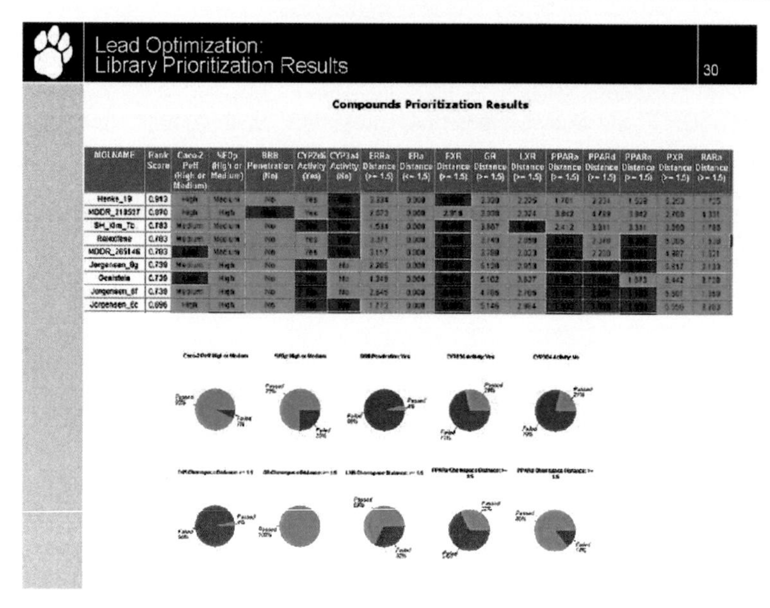

Figure 10. Output from compound prioritisation. Compound dependent rollup of predictions, and compound collection distributions of predictions.

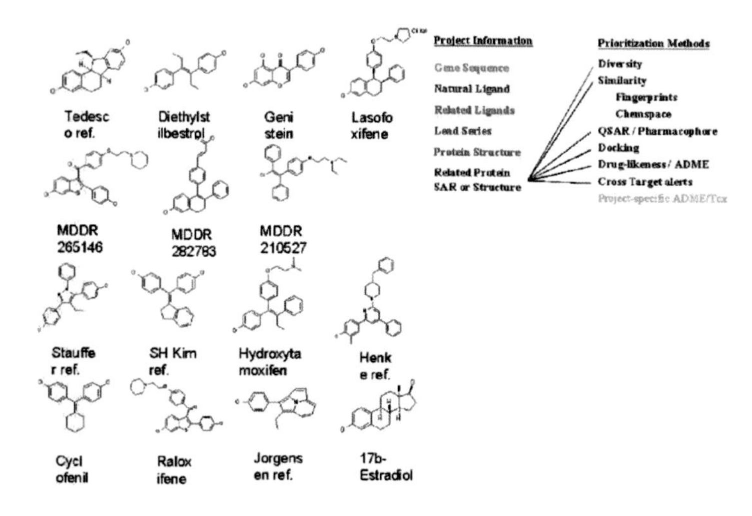

Figure 11. Starting information for the hypothetical ER-a project.

In addition, since our organisation is using a target class approach to drug discovery, we also retrieve all known NR-ligand pairs from the literature as an initial knowledge environment. ER-a is an example of a data-rich project (Figure 11), and therefore almost every method available for computational model generation would be at our disposal. Once the project criteria and starting assumptions (Example shown in Figure 12) have been established, lead identification

(Figure 3) can begin. A starting potential screening set is selected by a search of the known ligands against all available compounds (both virtual and existing - either in house or available from vendors - example shown in Figure 6).

This screening set is further reduced by simultaneous parameter evaluation using integrated computational tools. The project team decides where to cut off the screening collection, and these compounds are assembled, synthesised or acquired, and are run in initial potency assays.

	Cmpd ID	Activity		Selectivity			In Vitro ADME		In Vivo ADME		Score
Criteria / Rank		Primary Activity (EC50, uM)	Second Activity (%)	CR-1 (ratio)	CR-2 (ratio)	CR-3 (ratio)	Caco2 (Log Pe)	Hep. Turnover (%)	% F rat	%F hum.	Cum.
H		< 50 nM	>70	> 100	> 100	> 1000	< 5	> 70%	>50%	>50%	
M		50-500 nM	30-70	10-100	10-100	100-1000	5.-6.	30-70	10-50%	10-50%	
L		[illegible]	< 30	[illegible]	< 10	< 100	[illegible]	< 30	[illegible]	< 10	
Weight											

- Utilize "predicted" and / or determined parameters for compound design / prioritization and assessment / promotion
- Develop "Score" to roll-up key project criteria to assist decision making and project tracking

Figure 12. Project criteria and use example.

In this scenario, all evaluations can be performed by any and all members of the project team through a web site with common default settings for project criterion. After the screening results are returned and confirmed, the project team then must assess where the project is at that point in time, and decide where to focus the available resources. In this example, three series of compounds passed the requirements necessary for lead optimisation (Figure 13), however, the project team has only enough synthetic resources to work on one series at at time, therefore, the project team must decide which series has the most potential for success.

Lead Scaffolds:

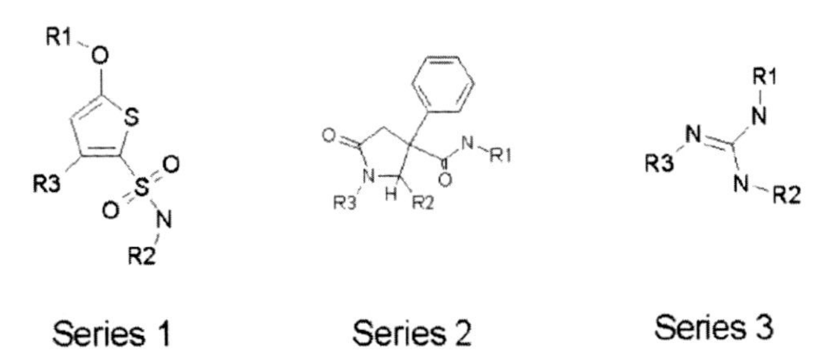

Figure 13. Hypothetical leads to choose between for prioritisation for optimisation.

To evaluate the potential of each series, virtual libraries will be enumerated, and the resultant products will be evaluated simultaneously (Figure 9). Then, to rate the potential of the libraries in the context of the other libraries, the distributions the components are compared simultaneously between the libraries representing the three scaffolds. Shown in Figure 14, this analysis can be used as an evaluation of potential liabilities which are best compared by analysing the fraction of each library that fails the project criteria for that component.

In this analysis, series 1 has the fewest bad distribution of liabilities, and should be chosen by the project team for further research, with series 2 representing a potential backup series.

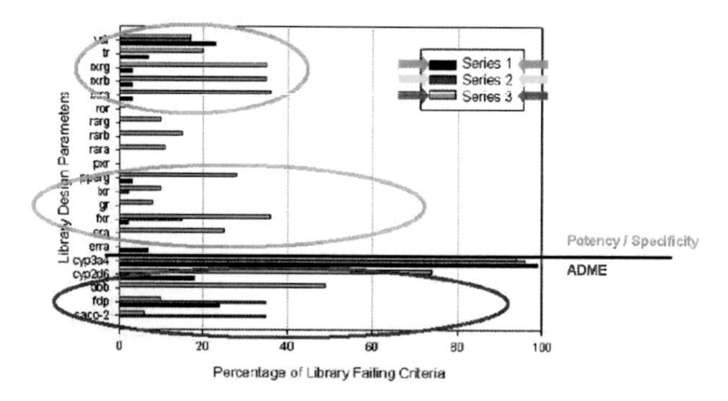

Figure 14. Simultaneous library property distribution comparison.

As the project moves closer to lead selection fpr pre-clinical development, the data integration components start to take priority over the in silico predictions. However, computational approches still have tremendous value at this stage, allowing the project team to evaluate the potential performance of candidate molecules in man. Figure 15 shows an analysis of the human absorption potential for three compounds representative of candidate series, in which the experimentally determined solubility and permeability values have been varied systematically.

As can be seen in the graphs, three classes of behaviors can be observed. In one example neither increases in solubility nor permeability can increase the absorption potential, which remains relatively low. In the second example, increases in solubility or permeabiltiy also have no effect on the human absorption potential, which is high. In the final example, increases in solubility and permeability show marked changes in the human absorption potential. The candidate molecules from the second and third examples are therefore the compound series to bring forward if all other factors besides absorption are equal. In addition, this analysis illustrates which series deserve followup study for the development of second generation compounds.

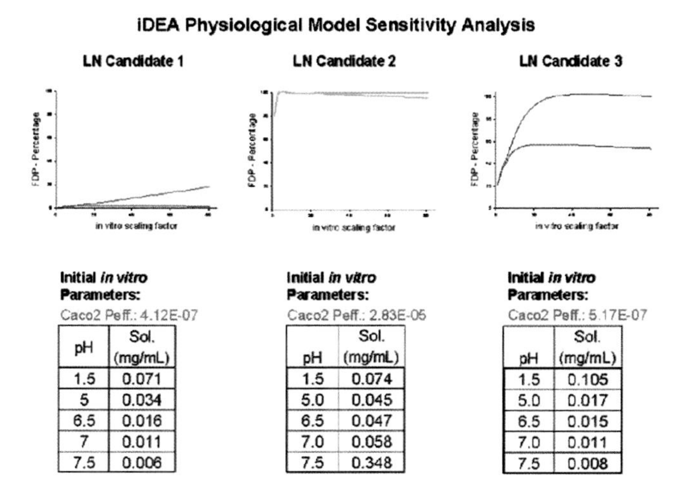

LN Candidate 1

Initial *in vitro* Parameters:
Caco2 Peff.: 4.12E-07

pH	Sol. (mg/mL)
1.5	0.071
5	0.034
6.5	0.016
7	0.011
7.5	0.006

LN Candidate 2

Initial *in vitro* Parameters:
Caco2 Peff.: 2.83E-05

pH	Sol. (mg/mL)
1.5	0.074
5.0	0.045
6.5	0.047
7.0	0.058
7.5	0.348

LN Candidate 3

Initial *in vitro* Parameters:
Caco2 Peff.: 5.17E-07

pH	Sol. (mg/mL)
1.5	0.105
5.0	0.017
6.5	0.015
7.0	0.011
7.5	0.008

Figure 15. Use of sensitivity analysis for the evaluation of human absorption potential for lead selection.

SUMMARY/CONCLUSION

So what is different about this approach? The drug discovery process is a fairly evolved one. Data has been shared and models developed for use in drug discovery process ever since affordable computers showed up in the marketplace.

However, the primary storage locale for experimental data is still Microsoft Excel, and the primary conduit of information is through individual interactions between project team members. Unfortunately, humans do not work with complete fidelity in serving as data or knowledge-sharing nodes. One could argue that modern drug discovery is using the ancient means of folklore as the method of knowledge sharing - this clearly is not sufficient given the ever exploding number of targets, hits, and interactions. This paper has described the initial steps of building a system that uses integrated databases to store *all* the data determined for discovery projects. It also indicates that this data is best used in synthesised from through the appropriate application of computational model building and their resultant use. Finally it illustrates the potentional power of combining the data and computational models in a system that allows the end user to simultaneously consider all available properties, and therefore make more and presumable better decisions about prioritising resources in the support of drug discovery and development.

DOES QUANTUM CHEMISTRY HAVE A PLACE IN CHEMINFORMATICS?

TIMOTHY CLARK

Computer-Chemie-Centrum, Friedrich-Alexander-Universität Erlangen-Nürnberg,
Nägelsbachstrasse 25, 91052 Erlangen, Germany.
E-Mail: clark@chemie.uni-erlangen.de

Received: 9th July 2002 / Published: 15th May 2003

ABSTRACT

The possible role of quantum mechanical (QM) techniques in cheminformatics is discussed. The advantages, disadvantages and capabilities of QM and its applicability to databases of thousands of molecules are discussed. The critical relationship between quantitative structure-property relationships (QSPRs) and the quality of the experimental data is discussed using aqueous solubility as an example. The use of QM-derived descriptors to investigate physical property space and to characterise compounds as drug-like or non-drug-like is illustrated. Finally, it is pointed out that not QM-calculations, but rather a knowledge of the molecular electron density is necessary for the examples shown, and a technique that can reproduce the electron density without QM-calculations is presented.

INTRODUCTION

Quantum mechanical calculations are not usually considered to be applicable to cheminformatics, although we have shown that semiempirical MO-calculations can be used on complete databases (1) and can play an important role in many cheminformatics applications (2,3). This article is intended to provide an overview of the applicability and capabilities of quantum mechanical techniques for cheminformatics and to discuss the relationships between data, descriptors and properties in quantitative structure-property relationships (QSPRs). Finally, an alternative technique for deriving the molecular electron density without quantum mechanics will be described.

Typically, cheminformatics applications use 2D- or very simple, classically derived 3D-descriptors for quantitative structure-activity relationships (QSARs) and QSPRs. As the border between, for instance, QSAR and pharmacophore-based high-throughput virtual screening is very poorly defined, many cheminformatics tasks can be considered to be simply more traditional QSAR or QSPR applied to larger numbers of molecules. In this respect, the constant advances in hard- and software performance tend to make the border even less clear because larger datasets become manageable with every advance. We have shown (2,3) that complete databases of tens of thousand of compounds can be treated with economical quantum mechanical techniques and have discussed the advantages of detailed quantum mechanical descriptions of molecules for QSPR (2) and QSAR (3). What, however, has changed since in the two years since references 2 and 3 were written? Is quantum mechanics still a useful tool for cheminformatics? Will it displace more traditional techniques? Are there alternatives?

Why Use Quantum Mechanics?

The advantages of using semimpirical MO-calculations to calculate molecular descriptors have been described before (2,3) and will only be outlined briefly here. The resolution of the molecular electrostatic properties is generally higher (i.e. atoms are not treated isotropically) in quantum mechanical calculations. This generally results in better descriptions of the molecular electrostatic potential in important regions of the molecular surface (i.e. where bonding interactions occur). Furthermore, electronic properties such as polarisability, ionisation potentials, electron affinities, dipole and higher multipole moments etc. and descriptors derived from them often prove to be very useful descriptors, especially in QSPR-applications. An example of the use of such descriptors is given in our recent work on the hydrogen-bond acceptor strengths of nitrogen heterocycles (4). Note however, that many of the properties listed above can be obtained from the electron density, so that efficient methods for generating an accurate electron density without quantum mechanics are potentially of great interest for cheminformatics.

However, for the moment we should assess the reliability of the most computationally economical for of molecular orbital theory, the modern semiempirical techniques MNDO (5), AM1 (6), and PM3 (7) for calculating the properties listed above. These techniques are parameterised to reproduce experimental heats of formation, molecular structures, dipole moments and ionisation potentials (from Koopmans' theorem). These properties (especially the dipole moment) ensure that the electron densities calculated are generally quite accurate, so that the molecular electrostatics calculated by semiempirical techniques agree well with high level *ab initio* data (8, 9). However, what about less directly parameterised properties like the polarisability, which is often thought to be very difficult to calculate and which requires very extensive basis sets at *ab initio* levels of theory? The solution, as for many properties in semiempirical theory, is to use a fast and effective level of theory to calculate the property in question and parameterise the method against experimental data. In this case, Rivail and his coworkers (10) published a very fast and simple variational technique for calculating the molecular electronic polarisability. However, this variational method is prone to systematic and element-specific errors, so that we (11) parameterised the element-specific integrals involved especially to reproduce the experimental data. This resulted in a fast and accurate method for calculation of the molecular electronic polarisability that is also amenable to partitioning into group, atomic or even orbital contributions (3, 12). The molecular electronic polarisability proves to be an important descriptor in most of our QSPR models and plays a very significant role in describing physical property space (13).

The general impression is that CPU-requirements for quantum mechanical calculations preclude them from being used for cheminformatics applications. This is not necessarily the case. In a first feasibility study (1), we were able to process the entire Maybridge database (about 53,000 compounds) on a 128-processor SGI Origin 2000 in half a day. However, computer performance has increased, and above all prices have decreased, since this study was performed, so that a Euro 1000 computer can process about 1,500 typical druglike compounds per day (full optimisations with AM1 or PM3).

Thus, there are relatively few real obstacles to using semiempirical MO-theory for cheminformatics. Do we, however, really need "better" descriptors, for instance for QSPR applications?

Experimental Data and QSPR Models

QSPR-models are derived by calculating descriptors for each molecule in the dataset and then using an interpolation technique (regression, neural net etc.) to relate the descriptors to the property. The quality of the model obtained is necessarily limited by the quality of the experimental data. Have we, however, already reached the limit of data-accuracy for some properties? Figure 1 shows results for an AM1/neural net model (14) for aqueous solubility based on a training set of solubilities for 559 compounds at 298K. This model gives a standard deviation between calculation and experiment of 0.51 log units, a mean unsigned error (MUE) of 0.40, a maximum error of 1.67 and 35% of the predictions outside the calculated (15) (± 1 standard deviation) error bars.

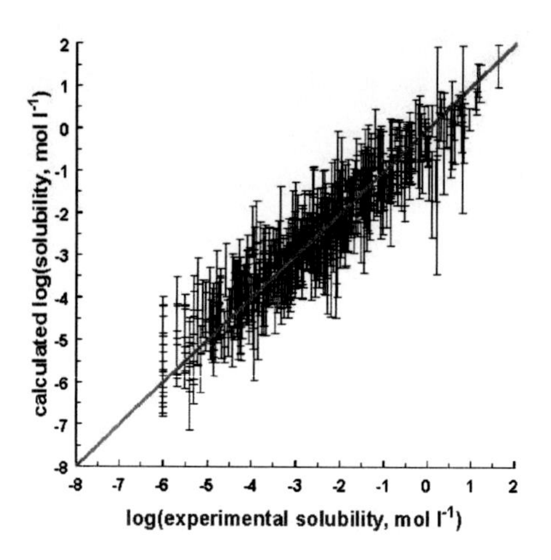

Figure 1. The performance of an artificial neural net QSPR-model[14] for aqueous solubility based on AM1 descriptors. The error bars shown are calculated according to the procedure outlined in reference 15.

These results are typical. A literature survey reveals that for 11 studies (some of which used related experimental data, not just calculated descriptors) using datasets between 399 and 1312 compounds, the six published standard deviations between calculation and experiment average to 0.60 log units (including one extreme outlier at 0.16) and the five published root mean square deviations (RMSD) average to 0.68. If we ignore the outlier, the standard deviations range from 0.57 to 0.79 and the RMSDs from 0.62 to 0.76. Thus, all published models perform very similarly, although they use very different types of descriptors and interpolation techniques. What is not different, however, are the data. How reliable are experimental solubility measurements? Yalkowsky and Banerjee[16] have outlined the experimental techniques for and difficulties encountered in measuring aqueous solubility. They also include a table of measured solubilities for some *"extremely hazardous substances"* (reference 16, Appendix C). As a crude estimate of the reliability of experimental data, Figure 2 shows the highest experimental value plotted against the lowest for the 18 compounds for which more than one measurement is listed.

The standard deviation between the two sets of experimental values is 0.79 log units, the MUE is 0.48, the RMSD 0.76 and the largest error 1.93. Even though a sample of only 18 compounds cannot be considered reliable, the conclusion seems clear that QSPR-models cannot be very much better than this correlation.

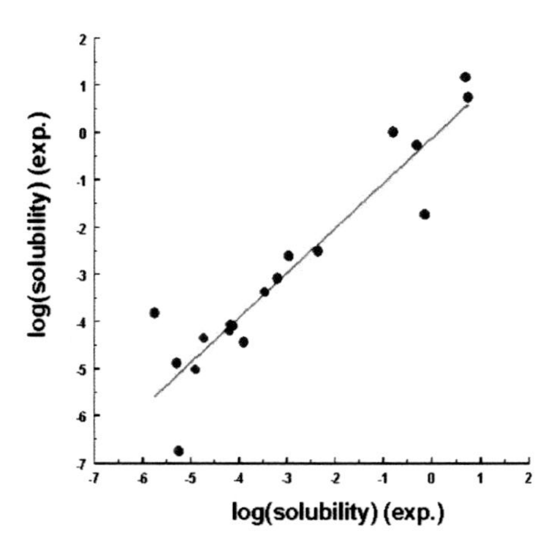

Figure 2. A plot of the highest experimental values for aqueous solubility against the lowest for 18 compounds taken from a table of *"extremely hazardous substances"* (reference 16, Appendix C).

We may even be in a situation in which our available descriptors could describe solubility very much better than they do if better experimental data were available (i.e. the models are limited by the data, not by the descriptors or the interpolation technique). Is it the worth developing new, "better" descriptors if the experimental data do not justify their use?

MAKING BETTER USE OF THE AVAILABLE DATA

It is not necessarily true that a QSPR model for a given property cannot be better than the dataset used to train for that property. Consider, for instance, molecular mechanics (force field) calculations for alkanes. Several high quality force fields give heats of formation for alkanes that are more reliable than experimentally measured values. This is possible because the force field (an unusual type of QSPR model) is not only trained to reproduce heats of formation, but also, for instance, structures, isomerisation energies etc. Thus, a very general QSPR model that is trained to reproduce several directly related properties can be more reliable than the experimental values for one or more of these properties. We recently tested (17) the applicability of a single QSPR model to more than one property for a simple example, vapour pressures. Note that we have not yet trained the model using data for more than one property, but have only tested the possibility that a single model can be used for several properties, in this case vapour pressure, boiling point and heat of vapourisation. Our first QSPR model for vapour pressure (15) used only data measured at 298K taken from the Beilstein database. This limits the training/test dataset to 551 compounds. Other authors (18) have corrected some of their experimental data to 298K using the published temperature dependence. This is a legitimate way to extend the available data. It is, however, not necessarily the best because it does not allow use of data for compounds for which the temperature dependence of the vapour pressure is not known.

Our approach (17) is to include the temperature of the measurement as an additional descriptor in the QSPR model. This forces the interpolation technique (in this case a feed forward neural net) to learn the temperature dependence as part of the model. We can then gain extra information by interrogating the trained net about this temperature dependence. In this case, the boiling point at atmospheric pressure can be calculated by finding the temperature at which the vapour pressure reaches this value and the heat of vapourisation can be derived using the Clausius-Clapeyron equation. Figure 3 gives an idea of exactly now much more data is available in this case than for a model limited to 298K.

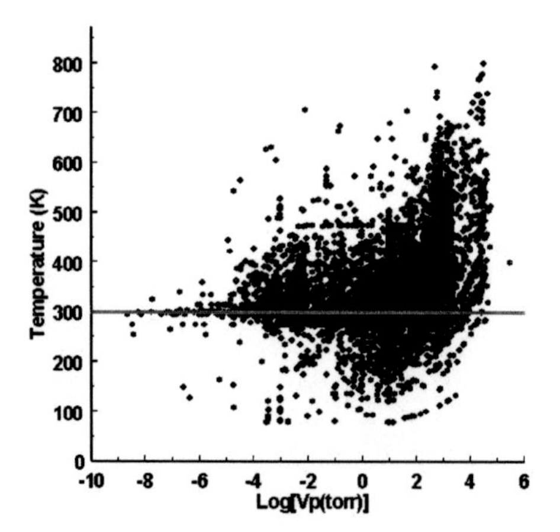

Figure 3. Distribution of the experimental data used to train a variable temperature vapour pressure model with respect to the vapour pressure itself and the temparature of the measurement. The horizontal line indicates the data available at 298 K.

Instead of the data for 551 compounds, we now have 8,542 data points at temperatures between 76K and 800K for 2,349 different compounds. Applying the model to boiling points gives a standard deviation between calculated values and experiment of 28.6K and a MUE of 18.7K for our boiling point dataset, compared with values of 21.9K and 13.5K, respectively, for a model

(19) trained only to reproduce boiling points. A small sample of heats of vaporisation also showed a standard deviation between calculation and experiment of only 4.7 kcal mol^{-1} and a maximum error of 12.0 kcal mol^{-1}.

This is, however, only an indication of what is possible. We must learn to make the most possible use of all the reliable data available. Martin Hicks has pointed out (20) that there are different types of experimental data. Boiling points, for instance, are measured routinely, and probably not very carefully, for every liquid organic compound reported in the literature. These data are not a good basis for a boiling point model because they are incidental properties used to characterise the compound. Our experience suggests that in many cases it is not recorded that the boiling point was measured at reduced pressure, or even that the temperature scale used (Celsius or Kelvin) is reported wrongly. There are, however, certainly data measured in studies whose main aim was to determine boiling points accurately. These data can all be used, even those at reduced pressure, because they are simply another way of reporting an experimental vapour pressure. Similarly, heats of vaporisation are usually measured very carefully for very pure compounds and so should be as reliable as is possible for measurements on a difficult quantity. Heats of vaporisation cannot be used directly to train a neural net by back-propagation because they require the slope of the vapour pressure change with temperature. However, training feed forward neural nets by genetic algorithms, rather than direct back-propagation, is an established technique (21) that allows us to use derived properties to determine the error function to be minimised as well as those generated directly by the neural net. We are now investigating the effectiveness of such an approach in which linked physical properties are used to generate more general, and thus more reliable, QSPR models (22).

Another potential use of information-rich molecular descriptors is to map chemical compounds according to their physical properties, as has been demonstrated by Oprea *et al.* (23) in their *"Chemical GPS"* technique. We have used this approach to investigate the clustering of druglike compounds on such a map, primarily in order to distinguish drugs from nondrugs, but as a more general goal to be able to relate new compounds to known closely related ones.

PHYSICAL PROPERTIES, DESCRIPTORS AND COMPOUND MAPS

The idea of mapping compounds according to, for instance, their physical properties is that physically similar but possibly chemically diverse compounds should occur close to each other, and thus have similar ADME properties etc. Thus, rather than conventional QSPR being used to predict individual properties, compounds would be compared with their known neighbours and assumed to behave similarly. So far so good, but how do we decide which descriptors (and how many) to use for the mapping? In order to be able to treat this question rationally, we should at least have some idea of the dimensionality and the descriptors appropriate for describing physical property space. Lipinksi (24) has described physical property space as being low dimensional (i.e. we only need a few descriptors to describe physical properties). We investigated (13) both the dimensionality and the nature of physical property space by calculating a range of descriptors known to be suitable for QSPR models for the entire Maybridge database plus a set of about 2,500 selected drugs. The principle components of the 26 descriptors that appear in many of our QSPR models were then calculated in order to characterise physical property space. The conclusions of this study are that 8-9 descriptors are enough to describe physical properties and that these can be loosely classified as shown in Table 1.

Table 1. Qualitative descriptions of the principle components of descriptors used to describe physical property space (13).

Principle component number	% variance explained	Main descriptors	Interpretation
1	23.3	Polarisability, molecular weight, surface area, globularity	Size, shape
2	18.5	Maximum MEP*, mean positive and negative MEPs, total variance (25)	Complementary electrostatic surface descriptors
3	9.1	Minimum MEP, mean negative MEP, balance parameter (25)	
4	7.6	Total MEP-derived charges on nitrogens (26), number of H-bond acceptors	Complementary hydrogen-bonding descriptors
5	5.4	Total MEP-derived charges on H and O (26), minimum MEP, number of aromatic rings	
6	5.4	Dipole moment, dipolar density (27)	Dipolar polarity
7-9	3.9 – 4.3	Total MEP-derived charges on different types of atoms	Chemical diversity

* MEP = molecular electrostatic potential at the solvent-excluded surface.

The interpretations of the individual PCs give an intuitive picture of the factors determining the physical properties of molecules. Most important are the size and shape, followed by two complementary descriptors that describe the higher multipole character of the electrostatics at the surface of the molecules. Note that the dipole moment is not important in these two descriptors. Next, come two complementary descriptors that essentially describe the hydrogen-bond donor and acceptor properties (including aromatic ring acceptors) and then the simple dipolar polarity, which perhaps surprisingly only accounts for just over 5% of the variance described by all the descriptors. PCs 7-9 are essentially atom counts that describe chemical diversity. These descriptors probably occur in QSPR models to correct for systematic AM1 or PM3 errors for some elements.

Thus, one appropriate approach to mapping chemicals according to their physical properties would be to use the first six principle components described in Table 1 as descriptors and to ignore the "chemical diversity" PCs in order to train, for instance, a Kohonen net. According to our analysis, the resulting map should cluster compounds with similar physical properties. This work is still in progress.

Descriptors can, however, also be selected to map for a more limited goal, such as distinguishing drugs from nondrugs (13). We have used recursive partitioning (28) In order to select descriptors for such a mapping, thus sacrificing some of the unsupervised quality of the Kohonen net by using a supervised descriptor selection process. The resulting map can not only distinguish drugs from nondrugs with about the same efficiency as other published techniques (29), but also differentiate between, for instance, hormones and other drugs (13).

The above applications use descriptors that are predominantly derived from the electron density, in our examples calculated using semiempirical MO-theory. However, it would be more efficient to use classical methods to optimise the molecular geometries and then a non-quantum technique to approximate the electron density. We are currently developing such a technique on which to base future, faster QSPR methods.

A Non Quantum Mechanical Approach to Electron Density

The principle of electronegativity equalisation (30) enjoyed some popularity 20 years ago and is the basis of the still popular Gasteiger-Marsili charges (31). However, all such models known to us calculate isotropic net atomic charges, rather than considering the inherent atomic anisotropies caused by the bonding situation. We are currently developing a procedure (32) that considers this atomic anisotropy by considering hybrid atomic orbitals and their interactions. Currently, we parameterise the model to reproduce the AM1 electron density, but high level *ab initio* or density functional data could also be used. Figure 4 shows a flow chart of the calculations steps involved.

The input to the program is a simple Lewis structure, which may be derived from a force-field calculation (in which case the bonds, hybridisation etc. are fully defined) or a set of 3D-coordinates, from which a Lewis structure must be derived using bond-distance criteria.

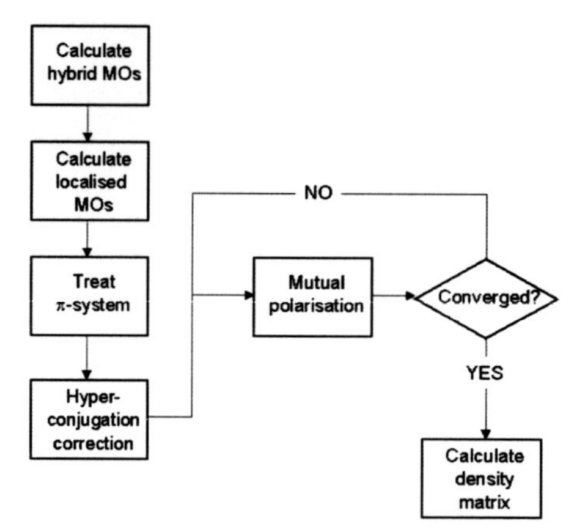

Figure 4. Flow chart of the classical procedure used to calculate electron densities (32).

The hybrid atomic orbitals are then calculated for each nonhydrogen atom using the procedure outlined previously for the hybrid orbital/point charge technique (33). The hybrid orbitals are then combined to bonding and antibonding localised molecular orbitals (LMOs) for the σ-framework and the lone pairs are identified. The remaining hybrid orbitals are considered components of the π-system, which is treated using a parameterised Hückel-like procedure with variable electronegativities. Negative hyperconjugation (donation from lone pairs into neighbouring σ*-LMOs) (34) corrections are added specifically at this stage. The σ-system is then allowed to undergo a mutual polarisation step analogous to electronegativity equalisation, but based on the electrostatic potentials at the nuclei (35). This is an iterative procedure that usually converges within 4-10 cycles.

The initial parameterisation was restricted to compounds of the elements H, C, N and O and without π-systems. A training set of 52 representative compounds was calculated with AM1 and the geometries thus obtained used for the parameterisation. The error function was based on the one-atom (4×4) blocks of the AM1 density matrix. The error in the off-diagonal elements was weighted with a factor of 0.5 compared with the diagonal elements. The error function was minimised with either the simplex or the BFGS optimisation algorithms.

A validation set of 25 cycloalkanes, ethers, amines, alcohols, sugars and steroids was used to test the resulting parameters. The preliminary results are shown in Table 2.

Table 2: Results obtained for the validation set of 25 compounds. "RMS" is the root mean square deviation between the target (AM1) value and that calculated by the non-quantum mechanical procedure.

Property	Numbers	RMS
Diagonal density matrix elements	1774	0.029
Off-diagonal one-atom density matrix elements	1986	0.029
Coulson atomic charges	781	0.037

The quality of the fit can also be expressed in properties that are more familiar. The steroid **1**, for instance, is part of the validation set.

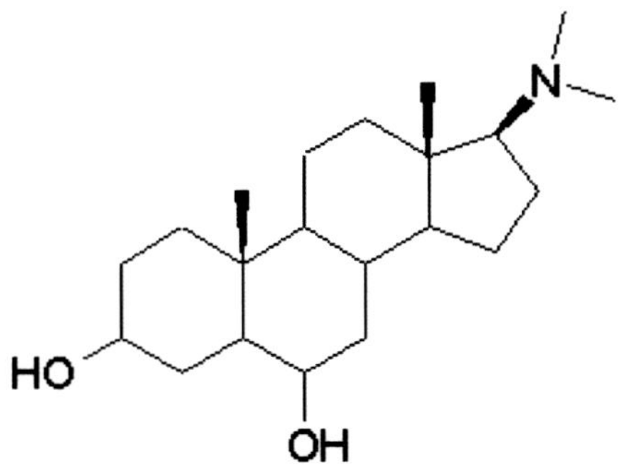

1

The parameterised procedure reproduces the AM1-calculated dipole moment with an error of 0.12 Debye, the root mean square of the Coulson net atomic charges is 0.027 and the root mean square deviation of the on-atom density matrix elements is 0.019. The molecular electrostatics of the molecule are thus well described by the new procedure, which requires only milliseconds of CPU-time. Figure 5 shows a plot of the molecular electrostatic potential at the solvent-excluded surface of the molecule. Only the areas with the largest deviation (below –5 kcal mol^{-1} and above 15 kcal mol^{-1}) are shown and the colour scale (blue to red) ranges from –9 kcal mol^{-1} to 19 kcal mol^{-1}.

The procedure outlined above would result in a vastly increased computational capacity for electron-density-based cheminformatics applications.

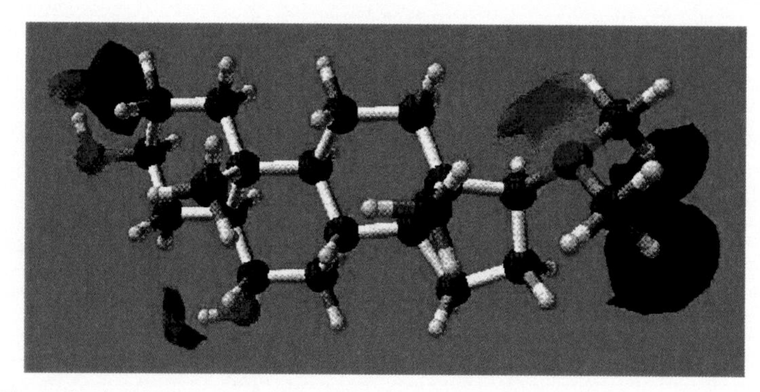

Figure 5. The molecular electrostatic potential at the solvent-excluded surface of **1**. Only the areas with the largest deviation (below –5 and above 15 kcal mol^{-1}) between the classical technique and the full AM1 calculation are shown. The colour scale (blue to red) ranges from –9 to 19 kcal mol^{-1}.

It is computationally very efficient, so that large databases or even complete enzymes can be treated easily. The electron density is also polarisable so that, for instance, the method could be used to calculate the electrostatics of the classical part of a hybrid QM/MM calculation without major inconsistencies in the electrostatic treatment of the classical and the quantum mechanical parts. Similarly, classically derived electron densities can be used as economical initial guess densities for MO-calculations. In this case, they have the advantage that no matrix diagonalisation is necessary, making the technique eminently suitable for parallel computers. A fast method for calculating accurate electron densities for proteins would also help the refinement of their X-ray structures.

CONCLUSIONS

Quantum mechanical methods, especially semiempirical MO-theory, can be used for cheminformatics applications. Advances in computer hardware have made semiempirical MO geometry optimisations on databases of 50-100,000 compounds commonplace on economical compute clusters. However, if we are to use the additional information provided by quantum mechanics relative to classical techniques, we must adopt a new paradigm for our QSPR and QSAR models, which are now often limited not by the descriptors, but rather by the quality of the training data. More general, physically rational models are needed that relate several physical properties to each other in order to eliminate biases or weaknesses in the training data for any one property. It is, for instance, unlikely that a dataset of heats of vapourisation will suffer from the same systematic problems as one for boiling points. High quality, possibly quantum mechanical descriptors will be needed should such compound QSPR models prove successful.

Many of our current descriptors, however, only require the electron density, not a wavefunction. An extension of the well known electronegativity equalisation technique to the calculations of a detailed electron density may prove to offer the ideal compromise between the detail offered by quantum mechanical calculations and the computational efficiency of classical methods. Our initial model has demonstrated the viability of such techniques. It promises to be of very general use wherever a fast, relatively accurate calculation of the electron density is required. As the algorithm is inherently parallel, it can be used for very large systems and may even be suitable for use in a polarisable force field.

ACKNOWLEDGEMENTS

This work was supported by the Fonds der Chemischen Industrie. I especially thank all my coworkers, who are named in the corresponding references and who have contributed enormously to the developments described above.

REFERENCES

[1] Beck, B., Horn, A., Carpenter, J. E., Clark, T. (1998). *J. Chem. Inf. Comput. Sci.* **38**:1214.

[2] *Quantum Cheminformatics: An Oxymoron?,* (Part 1) Published in *"Chemical Data Analysis in the Large: The Challenge of the Automation Age"*, M. G. Hicks (Ed.), *Proceedings of the Beilstein-Institut Workshop*, May 22nd - 26th, 2000, Bozen, Italy: http://www.beilstein-institut.de/bozen2000/proceedings

[3] *Quantum Cheminformatics: An Oxymoron?,* (Part 2) T. Clark (2001). In *Rational Approaches to Drug Design*, H.-D. Höltje & W. Sippl (Eds), Prous Science, Barcelona.

[4] Hennemann, M. & Clark, T. (2002). *J. Mol.Model.* **8**:95-101.

[5] Dewar, M. J. S. & Thiel, W. (1977). *J. Am. Chem. Soc.* **99**:4899, :4907; Thiel, W. (1998). *Encyclopedia of Computational Chemistry*, P. v. R. Schleyer, N. L. Allinger, T. Clark, J. Gasteiger, P. A. Kollman, H. F. Schaefer,III and P. R. Schreiner (Eds.), Wiley, Chichester, 1599.

[6] Dewar, M. J. S., Zoebisch, E. G., Healy, E. F., Stewart, J. J. P. (1985). *J. Am. Chem. Soc.* **107**:3902; Holder, A. J. (1998). *Encyclopedia of Computational Chemistry*, P. v. R. Schleyer, N. L. Allinger, T. Clark, J. Gasteiger, P. A. Kollman, H. F. Schaefer III, and P. R. Schreiner (Eds), Wiley, Chichester, 8.

[7] Stewart, J. J. P. (1989). *J. Comput. Chem.* **10**:209, :221; Stewart, J. J. P. (1998). *Encyclopedia of Computational Chemistry*, P. v. R. Schleyer, N. L. Allinger, T. Clark, J. Gasteiger, P. A. Kollman, H. F. Schaefer,III and P. R. Schreiner (Eds), Wiley, Chichester, 2080.

[8] Rauhut, G. & Clark, T. (1993). *J. Comput. Chem.* **14**:503.

[9] Beck, B., Rauhut, G., Clark, T. (1994). *J. Comput. Chem.* **15**:1064.

[10] Rinaldi, D. & Rivail, J.-L. (1974). *Theoret. Chim. Acta*, **32**:57, :243; Rivail, J.-L. & Carter, A. (1978). *Mol. Phys.* **36**:1085.

[11] Schürer, G, Gedeck, P., Gottschalk, M., Clark, T. (1999). *Int. J. Quant. Chem.* **75**:17.

[12] Martin, B., Gedeck, P., Clark, T. (2000). *Int. J. Quant. Chem.* **77**:473.

[13] Brüstle, M., Beck, B., Schindler, T., King, W., Mitchell, T., Clark, T. (2002). *J. Med. Chem.,* in press.

[14] Beck, B. & Clark, T., manuscript in preparation.

[15] Beck, B., Breindl, A., Clark, T. (2000). *J. Chem. Inf. Comput. Sci.* **40**:1046.

[16] Yalkowsky, S. H. & Banerjee, S. (1992). *Aqueous Solubility*, Marcel Dekker, New York.

[17] Chalk, A. J., Beck, B., Clark, T. (2001). *J. Chem. Inf. Comput. Sci.* **41**:1053.

[18] McClelland, H. E. & Jurs, P. C. (2000). *J. Chem. Inf. Comput. Sci.* **50**:967.

[19] Chalk, A. J., Beck, B., Clark, T. (2001). *J. Chem. Inf. Comput. Sci.* **41**:457.

[20] M. Hicks, personal communication and comment at the Bozen Workshop (2002).

[21] Montana, D. J. & Davis, L. D. (1989). In *Proceedings of the International Joint Conference on Artificial Intelligence*, Morgan Kaufmann, San Francisco; Mitchell, M. (1998). *An Introduction to genetic Algorithms*, The MIT Press, Cambridge, MA.

[22] Brüstle, M. & Clark, T., unpublished.

[23] Oprea, T. O. & Gottfries, J (2001). *ChemGPS: A Chemical Space Navigation Tool.* In *Rational Approaches to Drug Design: 13th European Symposium on QSAR*, H.-D. Höltje and W. Sippl (Eds), Prous Science, Barcelona, p. 437; Oprea, T. I. & Gottfries, J. (2001). *J. Comb. Chem.* **3**:157.

[24] Lipinski, C. A., Lombardo, F., Dominy, B. W., Feeney, P. J. (1997). *Avd. Drug. Delivery Rev.* **23**:3.

[25] Murray, J. S. & Politzer, P. (1998). *J. Mol. Struct. (Theochem)* **425**:107; Murray, J. S., Lane, P., Brinck, T., Paulsen, K., Grince, M. E., Politzer, P. (1993). *J. Phys. Chem.* **97**:9369.

[26] Beck, B., Clark, T., Glen, R. C. (1995). *J. Mol. Model.* **1**:176.

[27] Cronce, D. T., Famini, G. R., DeSoto, J. A., Wilson, L. Y. (1998). *J. Chem. Soc., Perkin Trans.* **2**:1293.

[28] Zhang, H. & Singer, B. (1999). *Recursive Partitioning in the Health Sciences*, Springer Verlag, Telos; Hawkins, D. M. http://www.stat.umn.edu/users/FIRM/.

[29] Sadowski, J. & Kubinyi, H. A. (1998). *J. Med. Chem.* **41**:3325; Wagener, M. & van Geersestein, V. J. (2000). *J. Chem. Inf. Comput. Sci.* **40**:280; Ajay; Walters, W. P. & Murcko, M. A. (1998). *J. Med. Chem.* **41**:3314.

[30] Sanderson, R. T. (1974). *Educ. Chem.* **11**:80.

[31] Gasteiger, J. & Marsili, M. (1978). *Tetrahedron Lett.* **34**:3181; Gasteiger, J. & Marsili, M. (1980). *Tetrahedron* **36**:3219; Marsilim, M. & Gasteiger, J. (1981). *Stud. Phys. Theor. Chem.* **16**:56; Marsili, M. & Gasteiger, J. (1981). *Croat. Chem. Acta* **53**:601.

[32] Horn, A. C. & Clark, T., unpublished.

[33] Gedeck, P., Schindler, T., Alex, A., Clark, T. (2000). *J. Mol. Model.* **6**:452.

[34] Schleyer, P. v. R. & Kos, A. J. (1983). *Tetrahedron* **39**:1141.

[35] Politzer, P. & Parr, R. G. (1974). *J. Chem. Phys.* **61**:4258; Politzer, P., Daiker, K. C., Trefonas, P., III. (1979). *J. Chem. Phys.* **70**:4400; Politzer, P. (1980). *Isr. J. Chem.* **19**:224; Politzer, P. & Sjoeberg, P. (1983). *J. Chem. Phys.* **78**:7008; Politzer, P. & Levy, M. (1987). *J. Chem. Phys.* **87**:5044; Erratum (1988). *J. Chem. Phys.* **89**:2590.

EPILOGUE - COMPLEXITY CHALLENGES RESEARCH IN MOLECULAR INFORMATICS

GISBERT SCHNEIDER

Beilstein Professor of Cheminformatics, Johann Wolfgang Goethe-Universität,
Institut für Organische Chemie und Chemische Biologie, Marie-Curie Str. 11,
D-60439 Frankfurt, Germany.
E-Mail: gisbert.schneider@modlab.de

Received: 20th Sept. 2002 / Published: 15th May 2003

"Molecular informatics" is a scientific discipline devoted to analysing and understanding the storage, processing and distribution of information encoded by molecules and molecular interactions, coined by contemporary bio- and cheminformatics research. Although this definition of molecular informatics may not be perfect, it is comparably easy to comprehend. The term "complexity" appears more vague and difficult to define. Although most of us do have an intuitive understanding of what complexity suggests, different persons will probably give a different answer to the question what complexity actually means and implies in the context of molecular informatics. The Beilstein-Workshop *Molecular Informatics: Confronting Complexity* held in Bozen, Italy, May 13-16, 2002, brought together an international group of scientists to present their research, exchange ideas and opinions, and discuss complex systems in the light of the workshop's challenging title.

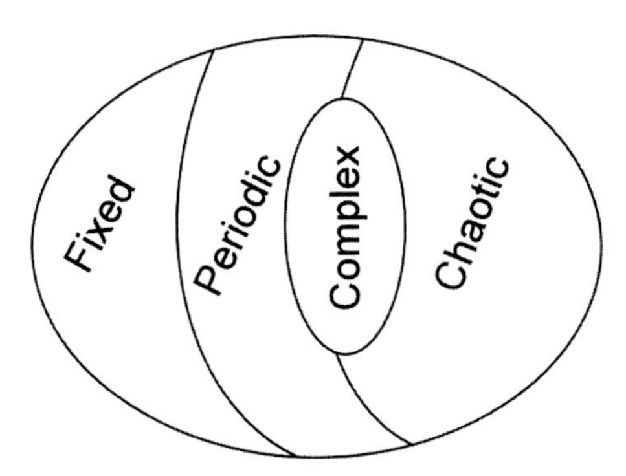

Figure 1. Complex systems may be placed between periodic and chaotic behavior. They show non-linear response, are partly unpredictable, and are characterized by the presence of noise. Graph adapted from ref. (1).

Complex systems may be characterized by three main attributes: i) partly unpredictable system behaviour, ii) non-linear response, and iii) inherent presence of noise. C. G. Langton located such systems "at the edge of chaos" (Figure 1) (1).

Typically, the objects of molecular informatics research are biological systems, like the structure and function of biological macromolecules, molecular recognition events, metabolic pathways and networks - all representing complex dynamical systems or their individual parts. It should be stressed that the term "complexity" is not the opposite of "simplicity". There are traditional scientific disciplines dealing with, e.g., algorithmic complexity addressing "orderly" systems that may be extended towards biological systems. On the other hand, a deeper understanding of biological complexity may be gained by methods such as advanced stochastic modelling and evolutionary computation, for which realizations on distributed computing facilities might be particularly well-suited (Figure 2).

The choice of methods and objects strongly depends on the scientific background and individual skills of a researcher, and several intriguing examples of both conceptual approaches are compiled in the workshop proceedings. An unifying theme during the workshop was the aim to gain insight into the behaviour of biological and molecular systems by computer simulation.

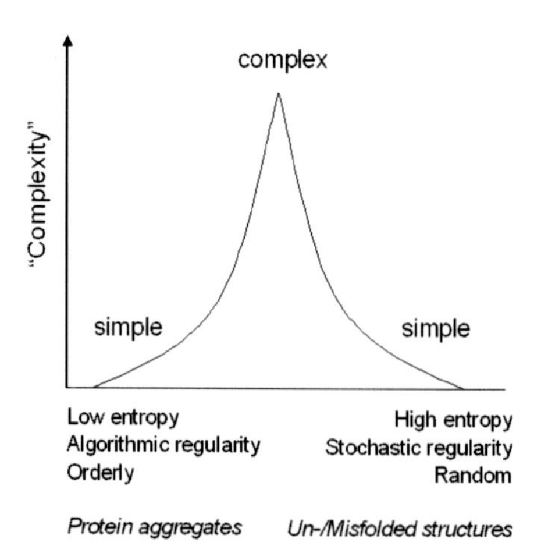

Figure 2. Complexity is not the opposite of simplicity. There are two types of problems generally regarded as tractable or "simple": orderly and random. Complex biological systems are neither entirely random or orderly, e.g. a protein's native state is neither an ordered aggregate nor unfolded. Graph adapted from ref. (6).

For example, realistic protein folding and molecular docking simulations are considered to be interrelated and represent very complex tasks. Approaches are derived from concepts abstracted from statistical mechanics, namely, populations, and from the purely physical standpoint, binding and folding are analogous processes, with similar underlying principles (2).

According to G. P. Williams there are six ingredients to complex dynamic systems (3):

1. A large number of items ("agents")
2. Dynamism
3. Adaptiveness
4. Self-organization (i.e. order forms inevitably or spontaneously)
5. Local rules that govern each agent
6. Hierarchical progression in the evolution of rules and structures

Confronted with this list, at the end of the Beilstein-workshop the participants were asked which of the six attributes of complex dynamic systems were best covered by the lectures. The result of this non-representative survey is summarized in Figure 3, revealing a clear trend.

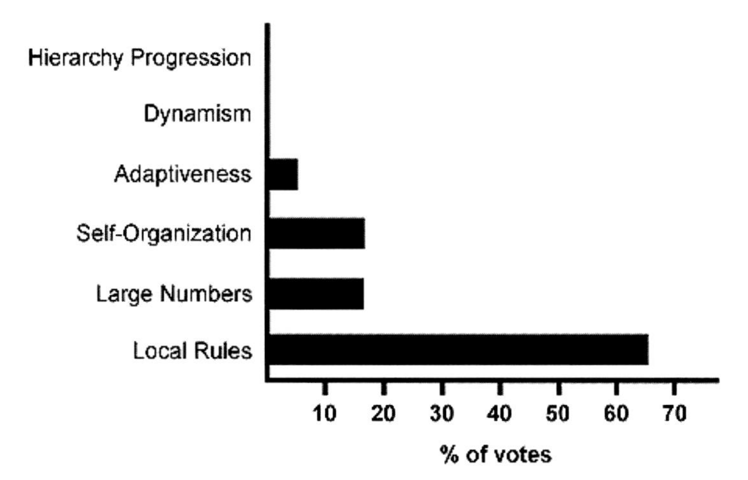

Figure 3. Result of a non-representative survey among the participants of the Beilstein-workshop 2002. The question was: "Which of the six attributes of complex dynamic systems was best covered by the lectures?"

Obviously the molecular informatics community seems to be rather familiar with the formulation of local rules and to some extent gives attention to issues related to self-optimisation and the problem of large numbers. At the same time it is obvious that essential attributes and properties of complex dynamic systems are not adequately or sufficiently treated by current molecular informatics research, namely the their dynamics, ability to adapt, and hierarchical evolution. Innovation is therefore needed to adequately treat other important attributes of complex biological systems. Generally innovation is considered to come in two equally important guises: i) unexpected, non-linear, quantum leap innovation; and ii) linear innovation based on incremental improvements (4,5). Appropriate working environments and conditions as well as the cross-fertilization of disciplines are needed for future success. It should be appreciated that the scientific community as a whole - and the group of workshop participants in particular - forms a complex dynamic system itself. As a consequence, there is good reason for hope that system-immanent mechanisms of development and optimisation will eventually lead to progress and innovation in the most challenging areas of molecular informatics research. The current research trends identified during the workshop are functional predictions in the field of genomics and proteomics, refinement of global approaches by modular rules, development of novel representations of biological and chemical information, and adaptation of methods from engineering, computer vision, and the machine learning community.

Acknowledgement

The author is grateful to the Beilstein-Institut zur Förderung der Chemischen Wissenschaften for generous support. Petra Schneider is thanked for valuable discussion during the preparation of the manuscript.

References

[1] Langton, C. G. (1991). Life at the edge of chaos. In: *Artificial Life II* (Langton, C.G., Taylor, C., Farmer, J.D., Rasmussen, S., Eds.). Addison-Wesley, Redwood City.

[2] Halperin, I., Ma, B., Wolfson, H., Nussinov, R. (2002). Principles of docking: An overview of search algorithms and a guide to scoring functions. *Proteins* **47**:409-443.

[3] Williams, G. P. (1997). *Chaos Theory Tamed.* Joseph Henry Press, Washington.

[4] Austin, A. (1998). Passion versus fear as the emotion driving scientists. *Drug Discov. Today* **3**:419-422.

[5] Schmid, E. F. (2002). Should scientific innovation be managed? *Drug Discov. Today* **7**:941-944.

[6] Flake, G. W. (1999). *The Computational Beauty of Nature.* M.I.T. Press, Cambridge.

BIOGRAPHIES

Karl-Heinz Baringhaus

was trained as an organic chemist at the University of Münster, Germany. After a postdoctoral at Stanford University he joined Hoechst AG in 1991. After six years within Medicinal Chemistry he moved 1997 into Molecular Modelling. In 1998 he became Head of Molecular Modeling and since 2000 he is coordinating global Aventis Pharma Molecular Modelling.

Tim Clark

was born in southern England and studied chemistry at the University of Kent at Canterbury, where he was awarded a first class honors Bachelor of Science degree in 1979. He obtained his Ph.D. from the Queen's University Belfast in 1973 after working on the thermochemistry and solid phase properties of adamantane and diamantane derivatives. After two years as an Imperial Chemical Industries Postdoctoral Fellow in Belfast, he moved to Princeton University as a NATO Postdoctoral Fellow working for Paul Schleyer in 1975. He then followed Schleyer to the Institut für Organische Chemie of the Universität Erlangen-Nürnberg in 1976. He is currently technical Director of the Comuter-Chemie-Centrum in Erlangen. His research areas include the development and application of quantum mechanical methods in inorganic, organic and biological chemistry, electron-transfer theory and the simulation of organic and inorganic reaction mechanisms. He is the author of 200 articles in scientific journals and two books and is the editor of the Journal of Molecular Modeling.

Athel Cornish-Bowden

carried out his undergraduate and post-graduate studies at Oxford, obtaining his D.Phil. with Jeremy R. Knowles in 1967 on the basis of studies of pepsin catalysis in the Dyson Perrins Laboratory. After spending three post-doctoral years in the laboratory of Daniel E. Koshland, Jr., at the University of California, Berkeley, he moved to a position as a Lecturer, and subsequently Senior Lecturer, in the Department of Biochemistry at the University of Birmingham, where he remained for 16 years. Since 1987 he has been Directeur de Recherche in three different laboratories of the CNRS at Marseilles. Despite having started his research career in a department of organic chemistry, essentially all of his research has been related to biochemistry in general and enzymes in particular, including pepsin, mammalian hexokinases, and bacterial enzymes involved in electron transfer. He is thus an enzymologist, with a major interest in kinetics, and has written several books in this area, including *Fundamentals of Enzyme Kinetics* (Portland Press, 1995) and *Analysis of Enzyme Kinetic Data* (Oxford University Press, 1995). In the past 15 years his interests have been focussed on multi-enzyme systems rather than on the kinetics of single enzymes. This topic includes the regulation of metabolic pathways, and his long-term aim is to develop a modern and coherent theory of metabolic regulation. There has always been a major element of computer analysis in his work, which at different times has involved statistical analysis of data, construction of protein phylogenies, and, most recently, modelling of metabolic systems.

Christopher M. Dobson

obtained his doctorate from the University of Oxford and has been assistant professor at Harvard University, visiting professor at MIT and from 1996 – 2001 Professor of Chemistry at Oxford University. In 1996, he was elected a Fellow of the Royal Society. In 2001, Chris Dobson became the John Humphrey Plummer Professor of Chemical and Structural Biology in the University of Cambridge. This post is a highly prestigious interdepartmental appointment with the objective of developing interdisciplinary research within Cambridge University. He is based in the Chemistry Department, but has joint appointments in the Departments of Physics and Biochemistry, as well as close links with the MRC Centre for Protein Engineering in Cambridge.

Throughout his career, Chris Dobson has been concerned with interdisciplinary approaches to understanding the nature and behaviour of proteins. His laboratory has utilised and developed a wide range of techniques based on NMR spectroscopy, mass spectrometry and electron microscopy, as well as a battery of biophysical techniques that utilise optical methods. His work has focused on using these techniques to understand the structural transitions involved in protein folding and to define the underlying mechanism of this complex process *in vitro* and *in vivo*. This work has recently led to novel insights into the phenomenon of protein misfolding and to studies of the conversion of normally soluble proteins into aggregated structures, including amyloid fibrils. His work links the prion diseases to fundamental events associated with protein folding and to other protein deposition diseases such as Alzheimer's disease and systemic non-neurological amyloidoses.

Chris Dobson has been distinguished by numerous international honours and awards and he has recently been awarded an Honorary Degree by the University of Leuven in Belgium. He will be Bakerian Lecturer of the Royal Society in 2003. He has published more than 350 papers over the last 20 years and has given a very wide range of lectures at research institutions and international meetings on the topics of his research.

Richard Goldstein

obtained his Ph.D. using experimental and computational methods to study electron transfer in bacterial photosythesis. After a brief stay teaching Physics in China, he worked with Peter Wolynes developing methods to predict protein tertiary structures. Since moving to a faculty position at the University of Michigan, he has worked on understanding the relationship between a protein's structure, function, and other properties and the evolutionary processes through which these properties emerged. These efforts have included methods of identifying and aligning distant protein homologs, examining the evolutionary record of related sets of proteins in order to determine characteristics of specific proteins, developing better models for phylogenetic reconstruction, and using simplified theoretical and computational models to develop deeper insights into the evolutionary process. Dr. Goldstein is in the process of leaving the University of Michigan and moving to Siena Biotech, a new Biotech company in Siena, Italy, where he will be leading the bioinformatics section.

Harren Jhoti

is Chief Scientific Officer and Founder of Astex Technology. He previously led the Structural Biology and Bioinformatics groups at Glaxo Wellcome (1991-1999), applying protein structure analysis to drug discovery. In addition, he was involved in structure-based drug design projects aimed at a variety of therapeutic targets including blood coagulation proteases, viral proteases, kinases and other signal transduction proteins. Dr Jhoti also played an active role in one of the BBSRC's research funding committees.

Dr. Jhoti graduated with a degree in Biochemistry from the University of London in 1985. He received his PhD in Protein Crystallography, in 1989, from Birkbeck College, University of London, UK. Dr Jhoti joined Glaxo in 1991 after completing a postdoctoral position investigating the structure of p21 ras, a potential cancer target, at Oxford University.

Carsten Kettner

studied biology at the University of Bonn and obtained his diploma at the University of Göttingen in the group of Prof. Gradmann which had the pioneering and futuristic name - "Molecular Electrobiology". This group consisted of people carrying out research in electrophysiology and molecular biology in fruitful cooperation. In this mixed environment, he studied transport characteristics of the yeast plasma membrane using patch clamp techniques. In 1996 he joined the group of Dr. Adam Bertl at the University of Karlsruhe and undertook research on another yeast membrane type. During this period, he successfully narrowed the gap between the biochemical and genetic properties, and the biophysical comprehension of the vacuolar proton-translocating ATP-hydrolase. He was awarded his Ph.D for this work in 1999. As a post-doctoral student he continued both the studies on the biophysical properties of the pump and investigated the kinetics and regulation of the dominant plasma membrane potassium channel (TOK1). In 2000 he moved to the Beilstein-Institut to represent the biological section of the funding department.

Christopher A. Lipinski

is a Senior Research Fellow in the Exploratory Medicinal Sciences Department at the Pfizer Global Research and Development Groton Laboratories. He received a B.Sc. degree in chemistry from San Francisco State College in 1965 and a Ph.D. in 1968 in physical organic chemistry from the University of California, Berkeley.

He joined Pfizer in 1970 following a National Institutes of General Medical Sciences Postdoctoral Fellowship at the California Institute of Technology. At Pfizer from 1970 to 1990 he supervised medicinal chemistry drug discovery laboratories discovering multiple gastrointestinal and diabetic clinical candidates. In this process, he became interested in the design of bioisosteres and in drug physical chemical properties and quantitative structure activity relationships, especially as they related to problems of oral activity. In 1990 he established a highly automated laboratory combining computations and experimental physical property measurements. Computationally he champions a very pragmatic, chemistry end user oriented, approach to the problem of oral activity improvement. Experimentally, his laboratory now provides experimental solubility measurements on medicinal compounds synthesized at the Pfizer Groton site.

He is a member of the Medicinal Chemistry section of the American Chemical Society, a member of the American Association of Pharmaceutical Sciences and a member of the Scientific Advisory Council of the Global Alliance for TB Drug Development.Since 1984, he has been an adjunct faculty member at Connecticut College in New London CT, and has over 120 publications, patents and invited presentations.

Gerald ("Gerry") M. Maggiora

Gerry received a B.S. in chemistry (1964) and a Ph.D. in biophysics (1968) from the University of California at Davis. He did postdoctoral work in theoretical chemistry at the University of Kansas and spent 15 years there as a faculty member in the Departments of Chemistry and Biochemistry. In 1985 he joinied what was then The Upjohn Company as the Director of Computational Chemistry, a position he held until 1998. Currently, Gerry is a Senior Research Scientist at Pharmacia Corporation.

During his tenure at the University of Kansas, he carried out research in the theory and application of quantum and molecular mechanical methods to problems in chemistry and biology. His interests included photosynthetic energy conversion, molecular spectroscopy, mechanisms of chemical and enzyme-catalyzed reactions, and protein structure and function.

Currently his interests include computer-aided drug design, protein-structure analysis and prediction, molecular similarity, thermodynamics of ligand-protein binding, biological systems theory, mining large-scale datasets, and applications of fuzzy mathematical and information theoretic methods to problems in chemistry and biology (with emphasis on applications relevant to drug discovery). During the 1994-95 academic year he completed a sabbatical at the University of New Mexico (1994-95) studying the theory of fuzzy mathematics and investigating selected applications.

Gerry has published more than 120 scientific papers and given numerous presentations at universities and at national and international meetings in the above scientifc areas. He also serves on a number of scientific advisory and editorial boards.

Luc De Raedt

received his undergraduate and Ph.D. degree in Computer Science (Licentiaat Informatica) from the Katholieke Universiteit Leuven (Belgium) in 1986 and 1991. He worked at the Department of Computer Science of the Katholieke Universiteit Leuven (Belgium) from 1986 till 1999, where he held positions as teaching assistant (1986-1991), post-doctoral researcher (1991-1999), part-time assistant professor (1993-1998) and part-time associate professor (1998-1999). From 1991 till 1999, he was a post-doctoral researcher of the Fund for Scientific Research, Flanders. Since April 1999 he is a full professor (C4) at the Albert-Ludwigs-Universitaet Freiburg and head of the Machine Learning and Natural Language Processing Lab research group.

Luc De Raedt's research interests are in Machine Learning and Data Mining and their applications in bio- and chemo-informatics. More recently, he has become interested in inductive databases, which are databases that allow one to store and query data as well as patterns. Luc De Raedt has been or is still involved in a number of European projects, such as ILP 1-2, Aladin, cInQ and April. He is on the editorial board of journals such as New Generation Computing, AI Communications, Intelligent Data Analysis, and Informatica, the Journal of Machine Learning Research. In 2001, he has organised and co-chaired the 5th European Conference on Principles and Practice of Knowledge Discovery in Databases and the 12th European Conference on Machine Learning in Freiburg. It was the first time – world-wide – that an important data mining conference was organised with one on machine learning. The conferences were attended by over 300 participants from all over the world.

Graham Richards

is chairman of chemistry at the University of Oxford. He is the author of over 350 papers and 15 books and was the scientific founder of Oxford Molecular Group Plc and more recently of Inhibox Ltd. He was awarded the Italgas Prize for 2001 for Computer-aided Drug Design and the Mullard award of The Royal Society in 1998. He is currently building a $100 million research laboratory for the Oxford Chemistry Department. The screen saver project which he devised started in April 2001 and now has over 1.5 million PCs involved, an effective teraflop machine which has provided over 100,000 years of CPU time and 215 countries.

Gisbert Schneider

born 1965 in Fulda, Germany; studied biochemistry and computer science at the Free University (FU) in Berlin; 1994, PhD in bioinformatics on neural networks and evolutionary algorithms; post-doctoral work on peptide design (with Prof. Wrede, FU Berlin), protein folding simulation (with Prof. Schimmel, M.I.T., Cambridge, USA), analysis of protein targeting signals (with Prof. von Heijne,University of Stockholm, Sweden) and prediction of membrane protein topology (with Prof. Schulten, Max-Planck-Institut Frankfurt, Germany); 1997-2002 F.Hoffmann-La Roche AG, Basel, Switzerland, head of cheminformatics; scientific research on combinatorial drug design, virtual screening, and genome analysis. Current position: Beilstein Professor of Cheminformatics at Johann Wolfgang Goethe-Universität, Frankfurt; research focus on adaptive systems in molecular design.

Brian Shoichet

attended college at MIT where he graduated with a B.Sc. degree in Chemistry and a B.Sc. degree in History in 1985. He then went to UCSF to study molecular docking and structure-based inhibitor discovery with Irwin Kuntz and received his Ph.D. at the end of 1991. After a one year postdoc with Kuntz, Shoichet went on to do postdoctoral research with Brian Matthews at the Institute of Molecular Biology in Eugene, OR, where he was a Damon Runyon Walter Winchel Cancer Research Fellow. There he studied protein structure and biophysics, focusing on a proposed balance between enzyme stability and activity. In 1996 he joined the faculty at Northwestern University as an Assistant Professor in the department of Molecular Pharmacology and Biological Chemistry.

At Northwestern, the Shoichet Laboratory remains interested in molecular docking, especially as it relates to inhibitor discovery. A focus is developing model systems for docking and inhibitor discovery, which has led to active research in the area of cavity sites and beta-lactamases. An interest in interpreting the structural "code" for molecular recognition has led to further investigations of how the stability of an enzyme constrains and identifies its function and its evolution. The laboratory remains actively involved in algorithm development for molecular docking.

Shoshana J. Wodak

received her Licence in Chemistry from the Université Libre de Bruxelles, this was followed by a Ph.D in Biophysics from Columbia University. Since 1981, she has held teaching and research positions at the Université Libre de Bruxelles, and in 1990 she became associate professor in the faculty of sciences. Between 1995-2002 she also held the position of Group Leader at the European Bioinformatics Inst. (EBI)

Shoshana Wodak has a wide range of research interests in structural computational biology and bioinformatics including: protein-protein and protein-nucleic acid interaction, protein structure and function, computational approaches to directed evolution, mapping protein-protein interactions onto cellular processes, and the development of databases and methods for representing and analyzing molecular and cellular function. Many software programs and tools have been developed in her laboratory, including the DESIGNER software for de-novo protein design and the aMAZE database for representing molecular function, interactions and cellular process. She has acted as a consultant and member of scientific advisory boards for a number of national and international companies, institutions and science foundations.

INDEX

INDEX

INDEX

INDEX

Index

INDEX

INDEX

S

T

INDEX

U / V

X

Y